DATE DUE

Bringing the Biosphere Home

Bringing the Biosphere Home

Learning to Perceive Global Environmental Change

Mitchell Thomashow

The MIT Press
Cambridge, Massachusetts
London, England

This book was set in Palatino by Achorn Graphic Services using QuarkXPress software and was printed on recycled paper and bound in the United States of America.

Library of Congress Cataloging-in-Publication Data

Thomashow, Mitchell.
  Bringing the biosphere home : learning to perceive global environmental change / Mitchell Thomashow.
    p. cm.
  Includes bibliographical references and index.
  ISBN 0-262-20137-2 (hc. : alk. paper)
  1. Global environmental change. I. Title.

GE149.T46 2001
363.738′74 — dc21   2001044033

To Jessica and Jake

Contents

Acknowledgments

Where do the origins of this book reside? At what point does an idea take hold? When does a project take over all aspects of your life? Perhaps some ideas are with you for your entire life, but fortuitous circumstances bring them to fruition.

For the last seven years I have been teaching Global Environmental Change in the Environmental Studies Program at Antioch New England Graduate School in Keene, New Hampshire. Much of what you read in *Bringing the Biosphere Home* emerged from the conversations, controversies, and inspirations that swirled through that class. My deepest thanks to the fine students I have the pleasure of working with.

Each year, as an advisory board member of The Orion Society, I attend the annual John Hay Award colloquium. I've had the great honor to visit with and celebrate some of the finest environmental writers and activists. Those meetings inevitably deal with issues of global environmental change, and I come away from them inspired and challenged. Without venues such as these, I would never have even known that this book could be written.

Hearty thanks to those people who have taken the time to read the manuscript in part or whole. I appreciate both your support and critiques: Beth Kaplin, Edward K. Kaplan, Bill McKibben, Lynn Margulis, Dorion Sagan, Scott Russell Sanders, Cynthia Thomashow, Mary Evelyn Tucker, Tyler Volk, Susan Ward, and Tom Wessels. I received terrific suggestions from The MIT Press "peer reviewers."

Thank you, Fran Silvestri, for the gift of your photography and friendship. I am honored that once again Lawrence Goldsmith has allowed me to choose a painting from his extraordinary collection. Larry is an exemplary biospheric artist.

Thanks to Frank Urbanowski and Clay Morgan at The MIT press. You both know how to put a book in perspective.

Chip Blake, the managing editor of *Orion Magazine,* provided terrific conceptual and editorial criticism. A writer could have no better friend, guide, and mentor.

Cynthia Thomashow, my spouse and colleague, provided ballast, common sense, perspective, and encouragement. Her love, support, and confidence are the source of so much inspiration.

In writing *Bringing the Biosphere Home,* I learned that its origins also lay in childhood memories, various travel experiences, and sources of learning that for many years had lain dormant. The issues in this book have always been with me. It's a book that I've been thinking about for as long as I have had conscious memories. So there is no limit to the extent of acknowledgments. And there is no closure either.

Yesterday, as I finished the final revisions, I followed the track of a good old-fashioned northeaster as it formed off the mid-Atlantic coast. Late in the day, as the storm was bearing down on New Hampshire, I went for a short cross-country ski run on the beaver pond across from the house. The snow was blowing horizontally and to return home I had to ski into the heart of the wind. I was only a quarter-mile from the house, yet I was in the grasp of the wild biosphere. A snowstorm brings you to the heart of mystery. What a fine reminder of the depth of one's unknowing.

December 31, 2000
Dublin, New Hampshire

Perhaps, fifty million years hence, members of some new species of animal, interested in paleoecology, will examine the geological record. They will note the transition that happened, from their point of view, fifty million years ago. Will they regard it as major or minor? Will it rate as the end of the Holocene epoch or even as the end of the Quaternary period? Or will it be no more than a minor episode early in the great Quaternary Glacial Age, during which some species (including Homo sapiens, perhaps?) went extinct and others shifted their ranges? We shall never know.

—A. C. Pielou, *After the Ice Age*

We would prefer to know. We prefer security. And the heart of responsibility is revealed when we see that faced with anxiety and the need for choice, precisely what we wish to know is that we cannot know.

—Herbert Fingarette, *The Self in Transformation*

Faith is the fruit of a seed planted in the depth of a lifetime.

—Abraham Joshua Heschel, *Man Is Not Alone*

The Perceptual Challenge of Global Environmental Change

Why I Am Writing This Book

On an unseasonably warm October morning, the hot sun lends a golden glow to the spectral New Hampshire foliage. The southwesterly breeze sends the first waves of falling leaves through the rich and pungent air. A damselfly lands on a bright orange maple leaf. A mosquito is perched on my thigh. Is this summer's last gasp or a harbinger of global climate change?

Later, at the dinner table, my adolescent son shows me the front page of the local newspaper, pointing out a colored computer map of El Niño, a glowing red pulsating image depicting the expanding boundaries of the warm ocean waters. The image represents a monstrous circulation pattern, as large as North America, that is interfering with Earth's complex weather systems. My son asks me whether El Niño is a cause for concern, if it means that the New England weather will change.

That evening, at a meeting of a local conservation group, community members wonder how to mobilize the town as to the ecological danger of invasive species. Purple loosestrife, a scourge of many New England wetlands, threatens to overwhelm several small ponds. This leads to a broader discussion about the global ecology of invasive species and how modern transportation allows seeds, spores, and critters to hitchhike across continents and oceans. We wonder whether any local action can possibly stem such a formidable trend.

Several weeks later, I attend a conference at the National Academy of Sciences in Washington, D.C. at which dozens of prominent conservation biologists, ecologists, and global change scientists gather to address the state of global biodiversity. The consensus is clear. The planet is on the verge of its sixth megaextinction, a devastating and irreparable loss to the intricate fabric of global ecosystems and the diversity of life.[1] For many of the participants, the conference unleashes a flood of emotional and spiritual anxiety, as well as a sense of urgency. How else do you respond to the litany of evidence regarding the decline of living systems and the uncertainties of environmental change?

It is not just a profound concern about the biodiversity crisis that unifies the scientists who speak at the conference. They also share similar approaches to perceiving the natural world. In almost every case, they participate in ongoing research about a local place, a specific habitat or species, or a series of questions about a population, community, or landscape. While engaged in research, they learn how to investigate that place in fine detail, using the best of their scientific expertise as well as whatever their five senses can observe. Before long, they come to know the place or species so well that they also develop an affiliation for it, a sense of what makes it unique and magnificent. Their sense of wonder is as strong as their abiding concern.

I am an environmental studies professor. I think, read, and teach about environmental issues all of the time. For twenty-five years, I have been working with people who are deeply interested in environmental questions and devote their lives and careers to addressing them. Despite all of this experience and attention, it is still hard for me (and my students and colleagues) to come to grips with the magnitude of global environmental change.

Why is that so? First, consider the sheer complexity of what's at stake. To conceptualize issues such as the loss of biodiversity and global climate change requires an understanding of ecology and evolution, an awareness of how the environment changes through geographic space and geological time. Second, these issues raise moral and existential dilemmas. If the environmental impact of human activities is profound enough to challenge the very fabric of ecosystems, what is the most appropriate ethical response? How do I assess my own actions and practices, or those of my neighbors, or those of my species in that regard? And if it is difficult for me, a person fully immersed in the environmental field, to consider these questions, then what must it feel like for the countless citizens who in one way or another have also shown some interest and concern, but who don't have the exposure and experience that I do?

Learning how to perceive the biosphere is crucial to understanding global environmental change. I use the term *biosphere* in its broadest and most literal sense, to convey the idea of a "sphere" of life ("bio") that surrounds the planet, its influence stretching from the highest reaches of the atmosphere to the inner depths of the earth's core.[2] This dynamic sphere coevolves with the planetary environment, creating a swirling biogeo-chemical playground whose elements combine to form patterns, cycles, and circulations of landscapes, species, and ecologies. Surely the idea of the biosphere is inconceivably complex. Yet you are born of the bio-sphere and will be buried in its earth. With every breath you take you participate in a dynamic exchange of global metabolism. We all respire together. By holding the idea of the biosphere in mind, in grappling with its complexity and magnificence, in aspiring to perceive its immanence, you experience the vitality of life and the very essence of biodiversity.

The biosphere comprises an entire matrix of planetary-scale ecologi-cal and evolutionary patterns, generically described as global environ-mental change.[3] I intend to show how the biosphere can be perceived if a person is interested in observing the natural world and willing to exper-iment with scale and perspective. To find the tracks, trails, and evidence of global environmental change you need look no further than your backyard, if you know what to observe. With study and practice, one learns how to connect what is close at hand to a whole globe of environ-mental changes.

Once the idea of global environmental change enters your awareness, what then? How do you continue to learn about it? To take responsible action? To fully reckon with the consequences? What happens thou-sands of miles away across the globe may dramatically affect your neighborhood. And the local development project just down the road from you may prompt a wave of ecological and political changes that will reverberate in communities you've never even heard of. Wherever you live, whatever you think about, developing an understanding of global environmental change dramatically expands your scope and vision.

As an environmental educator, global citizen, community member, and parent, it is my responsibility to reflect on how people learn about global environmental change. There are dozens of advocates who ring the bell to pay attention and take action. The register of environmen-tal threats is enough to intimidate anyone. But the role of educator is subtle—not just to raise awareness, but to encourage perception and facilitate wonder. Learning about global environmental change is an extraordinary opportunity to study natural history, observe ecosystems, track weather and atmospheric conditions, or to follow the cycling of water and energy through the land, air, and ocean. In so doing, the full splendor of the biosphere is revealed.

I am convinced that interpreting global environmental change re-quires an integration of three qualities—the best scientific thinking, the life of the imagination, and spiritual reflection. How else might one think about issues that pervade the entire planet? The biosphere is vast and mysterious. It permeates every pore of our bodies. The wind that rustles through the trees and cools my body on this hot, humid New Hampshire morning is a biospheric phenomenon, borne of air masses twirling around a spinning globe, churning energy, carrying seeds and spores. But it is tangible and visceral just the same. I can feel it move through my skin and bones. I listen to the sound of the rustling leaves. The wind is right here. But where does it come from? How far has it traveled? What atmospheric conditions created it? What message does it bring? The wind stirs my thinking just as it lifts my imagination and helps me soar beyond this place. As I reflect on the wind, I realize how the biosphere moves right through me. The wind is a means through which this place is connected to integrated biospheric cycles that circle the entire globe.[4]

I am writing this book because I want to bring the biosphere home. I want people to think about the various ways that global environmental change may be interpreted—how the threats and challenges of such problems as the loss of biodiversity, global climate change, and habitat degradation can become more accessible and personal, to the body and the mind, so they are directly perceived and intrinsic to everyday awareness. I am intrigued by the perceptual connections between the local and the global, how the ecological news of the community is of interest to the world, and how the global movement of peoples, species, and weather systems affects the local community. But most important, I am interested in how people learn about these connections and how global environmental change might become the province of countless educational initiatives—from the classroom to the Internet, from community forums to international conferences, from the backyard to the biosphere.

Wherever I move in various environmental circles—among scientists, policymakers, religious leaders, writers, artists, and educators—at conferences and gatherings, on the trail or in the classroom, I find that people are intrinsically concerned about global environmental change and crave ways to think about it. During chance encounters in airplanes, in supermarkets, at sporting events, or while waiting on lines, I often find a space to explore, with a friend or stranger, a mutual delight in the natural world—perhaps it's the weather, or the landscape, or wildlife. People are interested in understanding the remarkable environmental changes that surround them, and to do so through wonder and pleasure to balance their questions, anxieties, and concerns.

The Guiding Premises

At the core of this book is a simple, but counterintuitive educational assumption. The best way to learn how to perceive the biosphere is by paying close attention to the place where you live—developing familiarity and intimacy with local natural history. Behind this assumption is a deeply held value. Environmental learning must be built on a strong foundation in natural history and ecology. There is neither appreciation, awareness, nor affiliation with species and their habitats, without curiosity and concern about the natural world. There is no reduction of the extinction rate without a resurrection of such interest.[5]

This combination of assumption and value leads to four guiding premises. First I lay them out and then I briefly describe how they contribute to a vision for learning about global environmental change. The full measure of these premises will emerge as the book unfolds.

• Developing a place-based perceptual ecology is the foundation from which to interpret global environmental change.

• Once you are grounded in place, with a refined perceptual ecology, then you can learn how to move beyond that place and explore the relationship between places.

• By virtue of juxtaposing scale and perspective you learn how to explore the spatial and temporal dimensions of environmental change and thereby cultivate the ability to perceive the biosphere.

• You practice biospheric perception by virtue of three interconnected learning pathways—natural history and local ecology, the life of the imagination, and spiritual deliberation.

By place-based perceptual ecology I mean learning how to observe, witness, and interpret the ecological patterns of the place where you live. Why is a place-based orientation important? This is primarily a matter of scale and cognition. People are best equipped to observe what happens around them—what they can see, hear, smell, taste, and touch. These observations are poignant in their home places, where they are likely to spend lots of time, have many relationships, and be most in touch with the natural world. The home place is where you observe things closely, where you're most likely to develop significant affiliations. A place-based orientation is crucial to the contemporary nature writing tradition and at the forefront of current thinking in environmental education.[6]

I stress *perceptual ecology* because learning about landscapes, habitats, and species is both a perceptual and an ecological challenge, requiring specific observational skills and practices. To be a thoughtful naturalist means to acquire a range of facilities—the willingness to plunge your senses into the living landscape; the ability to ask good, scientific questions and develop approaches to finding empirical answers; the imaginative capacity to use the natural world as inspiration for artwork, photography, stories, essays, music, and poetry; the open-mindedness and reflective ability to be perennially engaged by the wonder, insight, and meaning derived from your observations. These ingredients together open the doors of perceptual ecology.

Once those doors are open, you have a solid foundation from which to move beyond your local observations. You can step back in time by learning how to perceive the tracks and trails of paleoenvironments. When you look at a boulder, you see a glacier. You come to know how various species arrive at your place, where they travel from, and how long they stay. When you see a scarlet tanager, you think about the tropics. You become more aware of interlocking cycles of change. From here in New Hampshire, clouds rolling in from the southwest are a way to trace the movement of weather systems—moisture from the Gulf of Mexico is traveling north. As you achieve facility observing environmental changes in space and time, you gain both visceral and conceptual glimpses of global processes and patterns.

Imagine the possibilities of biospheric perception. Next to your daily appointments calendar is a chart of the geological time scale. Next to the recycling rules posted on the refrigerator is a display of the carbon cycle. Adjacent to your family photographs is a five kingdoms taxonomic chart. You begin to perceive patterns of change that stretch from soil microorganisms to global energy budgets. You see the changing of the seasons as global respiration, the movement of weather patterns as global circulation, the movement of species as global migration.

You share your interpretations with local and global networks of similarly inclined observers. Perceiving the biosphere requires thousands of community eyes from outposts around the globe. People share their local natural history data, comparing trends and rates of change, wondering what patterns might emerge from their collective interpretations. Thousands of local natural history observers compare notes, sharing not just their data, but their artistic impressions too, building metaphors, constructing dialogues, talking about the moral and ethical implications of their observations. Schools, citizen groups, and clubs use the mass media and the Internet to broadcast the ecological news of the community. The patterns of global environmental change emerge seamlessly out of deep engagement with local natural history. Nodes of local observers form a global environmental change interpretive network—the biosphere observes and interprets itself.

The Meaning of Local and Global

A place-based orientation is not meant to suggest that one should only focus on local ecological concerns. Rather, the opposite is true. Local environmental decisions can be surprisingly elaborate and are rarely contained within their original boundaries. Think about any local environmental issue, say preserving a wetland, and you are sure to trace a chain of political causation that moves from your town to the state to national environmental policy, and perhaps to international development issues. The ecological considerations of that issue will touch on global questions such as invasive species, climate change, and biodiversity.

Global environmental change may be the invisible consequences of innumerable, seemingly unconnected local actions, spinning a synergy of effects and processes way beyond the original intention. Or it might be the extraordinary impact of one crucial choice or event. The more closely you look at any ecological or political controversy, no matter how tightly it seems to be bounded, the more you realize the extent to which the issue is informed and influenced by global patterns and processes. There is no such thing as a local environmental problem.[7]

It's been an entire generation, about thirty years, since I first heard the expression "think global, act local." At the time it seemed to make so much sense. I interpreted it to mean: participate in local decisions as if the fate of the earth is at stake. Deal with issues that can be locally resolved, but know that they are connected to a whole world of ecological and political change. There was an assumption that somehow one could make a global difference by attending to pressing local community issues. Inspired by photographs of the earth from space I tried to envision whole earth icons as the backdrop of all of my actions, everything from local political issues to the weekly trip to the recycling center. Over time I realized that this catchy slogan was far more complex than it initially appeared.

Through several decades of teaching environmental studies, I have found that the relationship between local and global is at the heart of personal and community concern. People want to feel that they can make a difference, that they can make both their community and the earth a better place to live in, that their actions have meaning. Yet they are typically overwhelmed by the magnitude of their charge. For the activist and policymaker, all kinds of strategic issues emerge—where can our actions have the most impact? The educator wonders how to approach teaching global environmental issues—how do you make such complex concepts pertinent and meaningful? For the scientist, local versus global patterns raises complicated questions of scale and perspective. How does your local research contribute to a body of global knowledge? And as concerned citizens, we all require guidance. How do you come to grips with environmental issues that are so sweeping as to contain the entire globe—climate change, biodiversity, and habitat degradation? Given the magnitude of these issues, what possible difference can your efforts make? The "perceptual challenge" of global environmental change lies in finding ways to allow these complicated, seemingly overwhelming issues to become tangible and accessible.

It's instructive to play with the permutations of the local/global slogan. One can act locally and think globally, or think locally and act globally, and pretty soon you come to realize that you're not even sure where you are any more. Peter Warshall, the editor of the *Whole Earth Review* has suggested the concept "globalocal" to indicate the interweaving of places.[8] Ronnie Lipschutz, a political scientist who writes about global environmental issues, reminds us that "everyone's experience of the world is centered where they are" and that "everyone is aware that the world is much more than the place in which they find themselves."[9]

Lipschutz challenges the distinction between local and global environmental issues:

To a major degree, this separation between 'global' and 'local' phenomena is the result of political boundary-drawing exercises, and not a consequence of nature. What, after all, makes powerplant pollution in one part of North America a 'transnational' problem even as, in another part of the same continent, not far away, it is a domestic one? Why should watershed degradation be regarded as a local problem if it is happening all over the world? Why, if toxic wastes are generated by an electronics firm whose markets are global, is their disposal strictly a local matter?[10]

What's crucial in learning to perceive global environmental change is that you practice how to think about the relationship between places, and in so doing, you come to understand how global patterns may be interpreted from the time and place where you happen to be. The most im-

portant perceptual skill is learning how to recognize the salient connections between seemingly disparate times and places.

From my rural outpost in the forested hills of southwest New Hampshire, in the course of all of my daily (local) routines, there are dozens of
reminders of the global context of my life. When I take my early morning
walk I plunge into a meteorological regime that covers several thousand
square miles. When I check my e-mail and glance at the newspaper
headlines online, I participate in an international communications network. As I sit down at my laptop and work away at this book, I am using
a machine that comes to me via a complicated, global production
process, involving dozens of far-ranging transactions. My morning meal
consists of locally grown strawberries sprinkled on organic cereal from
California. I may spend a week at home, working on this manuscript,
never getting in a car, taking lots of short walks, thinking about my
place-based life, writing a book about global environmental change. Is
the distinction between local and global merely a state of mind?

While I proclaim the great educational virtue of local observation,
singing its cognitive praises, I must be constantly vigilant against the
dangerous illusion that I can ever really understand this place without
traveling far afield. In the middle of working on this book, I had the opportunity to visit Mexico for the first time. After spending several unencumbered months in the rustic New Hampshire hills, there can be no
greater conceptual adjustment than flying into Mexico City. For miles on
end there is densely clustered human settlement, expanding without
limits—millions of people living on top of each other. Traveling west
from Mexico City, through central, urban Mexico I was amazed at the extent of industrial activity. And then the Mexican countryside presented
a dramatic contrast of stunning, lush mountainsides, and fast-paced
development, making its way up the steep terrain, leaving widespread
erosion in its wake. Here was a microcosm of global environmental
change—the pressures of population and poverty running smack
against the landscape.[11] It felt fraudulent to be writing a book about
global environmental change. What could I possibly understand about
the subject? How could I know anything, spending as much time as I do
in one small rural place in an affluent corner of the world?

My destination in Mexico was the wintering grounds of the monarch
butterfly. Every monarch east of the Rockies (see chapter 3) migrates to
the Oyamel fir forests of central Mexico which are now threatened by
deforestation. The tall fir trees are also a source of quick cash. In North

America, monarchs are also threatened, but for different reasons—the eradication and genetic manipulation of milkweed. Mega-agriculture, in its desire to support efficiency and remove unwanted plants ("weeds"), may also destroy the main food source for the monarchs. Different local circumstances are connected by a similar global process—the political economy of globalization emphasizes the efficient utilization of cash crop commodities.[12] Now, whenever I see a monarch butterfly in my forest garden, I know how far it has traveled to get here, and that its fate depends on the ecology of all the local places it encounters along its complex north-south journey. Previously I understood these concepts abstractly. Now I have the visceral experience and imagery to tighten the connections.

We all have a complex array of local places that we inhabit. What's crucial is how we compare them and which connections we take the time to explore. Your global vision will consist of whatever locals you choose to investigate. I advocate a place-based orientation as a means to closely observe what surrounds you, to realize the unfolding depths of space and time that any one place contains, to have a groundedness and foundation in the familiar, so when you explore other places you have a baseline for comparison. Sometimes there is a cognitive need for boundaries to interpret events and processes that can't really be contained. The distinction between local and global is helpful as a learning process. It helps you lay out what you can come to grips with, thereby developing a reasonable scope for your efforts and aspirations. It's a way to gain a visceral understanding of the idea that what happens on a planetary scale is also taking place right under your nose.

This is the crux of the matter—how to facilitate learning about global environmental change so that more people can observe that planetary scale processes like species extinction and global warming are happening in their local communities. That's where the consideration of expressions like local and global should lead us.

Why the Sixth Megaextinction Is Relatively Unknown

During the course of working on this book, people would ask me what I was writing about. Such questions are great opportunities to figure out what you think is most important about your work. After fumbling with inadequate responses for a while, one of the driving questions of the book became clear—how is it that we're on the verge of the sixth megaextinction and so few people seem to know or care? Quickly

I came to realize that the question required further explanation. Outside of the environmental community (and even within it sometimes) the expression "sixth megaextinction" drew blanks or politely raised eyebrows.[13]

I knew I had to explain myself simply and clearly, without judgment or jargon. There was no easy way to accomplish this. I would briefly describe how during the course of the history of life on earth there were five previous "catastrophic" extinctions, attributed to various geological, astronomical, and climatological causes. The great majority of conservation biologists agree that we are in the initial stages of a sixth one, with the primary difference being it's anthropogenic (human-induced) origins. This is not exactly grounds for casual conversation or a way to make new friends. Having established the "bad news" I would quickly move to another aspect of my work, my interest in using increased awareness of one's home place as a means to better understand and perceive global environmental change. I would gauge a more enthusiastic response, and my conversation partners would inevitably have something to add, as this topic came within the realm of their daily observations.

People would talk to me not only about their home place but also about their travels, trying to make sense of the rapid economic and environmental changes they'd observed and what it meant to them. A local businessman told me about his recent trip to China, and how astonished he was at the amount of commerce he saw. He vividly described roads lined for miles with trucks, carrying all sorts of goods to places he'd never heard of. He wondered about the scope of this activity, what its impact would be on energy use, pollution, and global warming. As a businessman, he was well aware of the driving force of commerce and understood why China and East Asia are moving so quickly to develop their industry and agriculture. He was aware as well, that even though he ran a relatively small business in a modest New Hampshire city, he was participating in a global economic process that had great impact on the place where he lives. Yet he had never heard the terms "sixth megaextinction" or "biodiversity." What's more, he doubted his ability to understand the terms or even to read further about them, worrying that the material would be too technical.[14]

After dozens of such conversations I realized that it was easy enough to get people interested in biodiversity and global environmental change, and that through their daily experiences they had been thinking about these issues in their own way. But there was still the sense that they were overwhelmed by the magnitude of the ideas. This is exactly the problem that people in the environmental field often face. How do you translate one's casual concern into a focused urgency?

I still trust my educator's instincts. By getting people involved in local natural history, you can awaken their interest in global environmental issues. An appreciation for the biosphere and biodiversity may accompany such interest. But I realized there was something much deeper at work. Awareness of biodiversity and megaextinctions transcends ecological and political considerations and opens you up to all kinds of existential dilemmas. Anyone who studies global environmental change will eventually confront profound spiritual challenges—the contemplation of love and loss, life and death, creation and extinction. Are environmental educators and activists even remotely prepared to raise these questions? "Bringing the biosphere home" may be just too much to take on. Perhaps the sixth megaextinction is relatively unknown, not only because it is too technical or that it receives far too little media coverage,[15] but because so few people are willing to confront the magnitude of the crisis.

Around the same time I began noticing the extraordinary increase in conferences dealing with religion and ecology. Many theologians, clergy, psychologists, and religious educators observed that planetary-scale environmental issues were related to the spiritual angst of their communities. Similarly, ecologists, conservation biologists, global change scientists, and environmental educators understood that they were confronting moral and ethical issues that required much wider discussion. It became clear that both groups needed each other to harbor the anxiety of their common concerns and to strategize the necessary education and advocacy.[16]

What's behind this interesting confluence of religion and ecology? There is a common understanding that interpreting global environmental change is an existential challenge raising questions of human purpose and meaning. For example, consider some of the scalar perspectives you might encounter in thinking about the biodiversity crisis. First, to place the issue in a biospheric context, you have to expand your sense of time so that you think in terms of the history of life on earth (four billion years or so). The idea of six "megaextinctions" cannot be fathomed without this deep time perspective. Second, you become aware of the relationship between creation and extinction, the vast number of species that have come and gone, the startling array of biogeochemical circumstances that make up ecology and evolution. Biodiversity doesn't take on meaning until one can appreciate what goes into it.

These perspectives are more than scientific conceptual challenges. They inform how you think about the human condition. Bringing the biosphere home may start simply enough as following a weather system across the continent to your backyard, continuing through a tour of Pleistocene climatic regimes, finally leading to a study of continental drift and the movements of land and ocean. How else do you get the big picture regarding global climate change? Learning to perceive biodiversity in your community may start with some solid natural history and ecology, looking at pond samples through a hand lens, and lead to broader questions about habitat change, speciation, and evolution. Given the vastness of these journeys, eventually you will come to contemplate where humans fit in the picture. Of what significance is your life, your community, your culture, or any actions that you take and thoughts you have? Do you respond to such questions with awe and wonder or with a sense of foreboding? Or perhaps with an intriguing mix of both? It makes sense that so few people are aware of or choose to think about the sixth megaextinction. It's really hard to do.

The perceptual challenge of global environmental change is an unprecedented and exhilarating educational opportunity. Consider what's involved:

- Taking the time to learn local natural history and ecology
- Making the connections between your daily routines and the global political economy
- Learning how to perceive the patterns of global climate change, biodiversity, species loss, and habitat transformation in your community
- Exploring the vast reaches of biospheric space and time
- Contemplating the meaning and purpose of human action

It's no surprise that conservation biologists, ecologists, global change scientists, environmental educators, theologians, psychologists, philosophers, and all kinds of people in between should strive to travel beyond their disciplines, forge new networks, and build coalitions. Learning to perceive and interpret global environmental change requires that you cover such wide-ranging territory. You need not be an expert or an academic to do so. Rather it takes a commitment to pay close attention to the intricacies of natural history, a concern and curiosity about the fate of the earth, and the willingness to confront the existential tensions that you will undoubtedly face, acknowledging the profound challenges, but in a way that prompts action and participation rather than despair and confusion.

The Plan of the Book in Brief

Bringing the Biosphere Home is organized so as to explore these ecological and existential challenges simultaneously. Natural history anecdotes and experiences are sprinkled throughout the text, linked with ideas for observational practice, and then used as the basis for broader contemplation. This book emphasizes perception and interpretation—learning to observe global environmental change is the first step, how you derive meaning from it is the second, and what you do about it is the third. Nevertheless, it's not a handbook, or a field guide, or an ecopolitical manifesto. It's a series of interconnected essays that explore various approaches to learning about global environmental change.

The next two chapters look more specifically at the perceptual challenges that I've introduced here. Chapter 2, "The Experience of Globality," explores the narrative and metaphorical context of global awareness. Over the last fifty years, by virtue of globalization, nuclear weaponry, whole earth imagery, and mass communications, various global images and metaphors have become increasingly prominent. These developments provide the setting for how people imagine and experience "globality." Yet one's awareness of a global economy or participation in a global communications network doesn't necessarily increase awareness of global environmental change. Instead, ideas such as species extinction and global warming pervade public awareness as a vague foreboding, a diffuse anxiety. What are the educational implications of this? Chapter 3, "Keeping Global Change in Mind," considers the existential challenges intrinsic to perceiving global environmental change. I describe several tensions that consistently emerge—creation and extinction, wonder and indifference, hope and foreboding. The chapter concludes with a discussion of how wonder spawns indebtedness and how biospheric perception might be considered an apprenticeship with hope.

Using the educational foundations of place-based environmental learning, Chapters 4 through 6 suggest means for cultivating biospheric perception. Chapter 4, "A Place-Based Perceptual Ecology," provides a series of observational approaches, derived from the naturalist sensibility, which serve to enhance one's ability to explore scale and perception. Covering the relationship between landscape and life cycle, how landscapes move through time, how to move between places, and exploring edges, the emphasis is on cognitive approaches to enhancing natural history observation. Chapter 5, "Interpreting the Biosphere," builds on these approaches as a means for further understanding environmental change. I show how "exemplary biospheric naturalists" rely on three pillars of inquiry—analysis, compassion, and imagination—to expand their perceptual and cognitive scope. Interpreting the biosphere involves tracking the four elements, tracking ancestry and lineage through deep time, and studying biospheric knowledge systems. Chapter 6, "The Internet, the Interstate, and the Biosphere," considers how the use of electronic communications and high-speed transportation both enhance and diminish environmental perception. Every time you use the Internet or get in an airplane you participate in a profound perceptual experiment that alters your perception of nature. Just as these technologies dramatically expand your perceptual vistas, they also accelerate your pace, making it difficult to observe the proximate details of intimate place-based learning. What is an appropriate perceptual balance for interpreting the biosphere?

While observing environmental change, one is struck by the fleeting quality of many ecological and human communities. Patterns of species migrations, human diasporas, and shifting landscape challenge the validity of a place-based orientation. People and species are constantly on the move. Chapter 7, "Place-Based Transience," suggests that migration and diaspora are a means to bring a global perspective to local ecology and community. How might the idea of transience broaden your scale and perspective, leading to a greater appreciation of ecological and cultural diversity?

Chapter 8, "A Biospheric Curriculum," delves more deeply into one of the primary cognitive approaches of the book, the emphasis on learning how to explore different scales of perception as a means to move between conceptual worlds. I suggest that the prefix "inter-" serve as the linguistic inspiration for this challenge, and that concepts such as interspatial, interspecies, intertemporal, and intergenerational may be the basis for biospheric pattern learning. This template might serve curriculum designers and educational researchers alike—we know very little about the cognitive dimensions of ecological learning and still less about biospheric perception. The chapter outlines some of the educational challenges and opportunities demanded by a biospheric curriculum.

Biospheric Perception Is a Practice

As an educator, I think a great deal about how people learn and what prevents them from learning. I'm especially intrigued at this question in regard to lifelong learning pursuits such as playing a musical instrument, meditation, or natural history. In this book, I'm interested in how people learn to perceive global environmental change, surely a lifelong learning commitment. Like any other endeavor, this requires patience, perseverance, and mentoring, as well as a sense of critical self-reflection. "Biospheric perception" is a practice. It's something you learn how to do. You don't become proficient as a jazz musician overnight. Nor should you expect biospheric perception to come easily either. Just as an aspiring jazz musician, for example, has to overcome all kinds of distractions to pursue a daily practice, so one has to achieve the same discipline in observing the natural world. And just as the musician coordinates structure (learning scales), improvisation (playing on the spot), and theory (understanding the relationship between the notes), so does the biospheric perceiver need to coordinate specific observational methods with an "improvisational gaze" and sound, fundamental global change science texts.

However, unlike music in which people are born with different proclivities, everyone can learn how to perceive the biosphere. As a living organism, it is your birthright. As a conscious human being, you have a chance to interpret your life from a biospheric perspective. Awareness of seemingly grandiose concepts like the biosphere and biodiversity are a means not only to reflect on the human condition, spiritual matters, and aesthetics, but they are the very basis of human survival.

This book is designed so as to provide some guidelines for learning how to practice biospheric perception. The idea of "bringing the biosphere home" is both metaphorical and literal. All my experience as an educator tells me that the reexamination of the routines and habits of daily life is the best way to immerse yourself in new ideas. I'd like to briefly lay out my approach so you know both what to expect and what I hope you'll spend some time thinking about.

Here are the methods I've taken both in writing the book and in working with students for the purposes of practicing biospheric perception. First, I emphasize the importance of the routine experience. In the course of your daily affairs and adventures, you have all the material you need for interpreting global environmental change. Biospheric perception is a practice you can engage in wherever you may be. In the time and space between your busy tasks, you can take a few moments to reconsider where you are, have a look around, and notice the sky, the landscape, and other life forms. In just a few thought moments you can travel a considerable conceptual distance through the biosphere. Second, I accentuate the narrative experience. I probe the stories that emerge from childhood memories, travels, and conversations, in conjunction with imaginative forays. To perceive the biosphere requires comparing times and places, different views you've had of the same spot through many years, understanding how your perceptions change by presence or absence. Imagination and memory often work together to conjure impressions that you may not attain in any other way. Third, I encourage you to carefully observe what you observe—knowing your proclivities and interests, assessing your insights, figuring out your perceptual and ecological strengths and weaknesses, the things that you see as well as the gaps, and using good teachers to help you in this. Fourth, biospheric perception is a community practice, something that you engage in with other people. It takes lots of folks pointing things out to each other to reap the deepest insights. Fifth, I emphasize the importance of global change science as a means to provide balance and ballast for your observations. The biosphere is not necessarily what you project it to be. It involves processes and patterns that are empirically derived.

Finally, I wind through a shifting phenomenological and existential passage. By phenomenology I refer to the great insights that can be derived from one's direct sensory impressions. To practice biospheric perception you must aspire to probe the full potential of your sensory awareness.[17] By existential I convey the impression that we are investigating ideas and concepts that we can never fully understand. There is a measure of cosmic frustration built into this work, a constant tension balancing the significance of your actions and their utter meaninglessness.

There are many times during the course of this work when one feels overwhelmed by the magnitude of the challenge. The prospects for reversing the sixth megaextinction are not good. The thrust of global resource extraction is just too rapid. The reawakening of public awareness that's required often seems impossible. Confronted with the outrageous world poverty figures, the egregious inequalities of the distribution of global wealth, and the unfettered quest for consumer satisfaction, it is easy to feel resigned and depressed. The reflective observer also has to come to grips with the stark reality of his or her culpability. These issues don't lie outside us in some far-flung corner of the world. Most readers of this book dwell in the belly of the beast. So acknowledge these impressions and respect them—one cannot think about global environmental change without encountering this reality.

Yet you don't have to be optimistic to be hopeful. You can't predict the future by virtue of a trend that you sit squarely in the middle of. And you can never assess the long-term impact of your thoughts and actions. The kabbalah suggests that when the universe was created, it was sprinkled with infinite sparks, each representing a microcosm or kernel of cosmic energy. The task for humanity is to search for and kindle those sparks.

We constantly aspire to raise the holy sparks. We know that the potent energy of the divine ideal—the splendor at the root of existence—has not yet been revealed and actualized in the world around us. Yet the entire momentum of being approaches that ideal.

The ideal ripens within our spirit as we ascend. As we become aware of the ideal, absorbing it from the abundance beyond bounded existence, we revive and restore all the fragments that we gather from life—from every motion, every force, every dealing, every sensation, every substance, trivial or vital. The scattered light stammers in the entirety, mouthing solitary syllables that combine into a dynamic song of creation. Sprinkling, the flowing, this light of life is suffused with holy energy.

We raise these scattered sparks and arrange them into worlds, constructed within us, in our private and social lives. In proportion to the sparks we raise, our lives are enriched. Everything accords with how we act. The higher the aspiration, the greater the action; the deeper the insight, the higher the aspiration.[18]

Perhaps learning how to interpret global environmental change is akin to gathering sparks. Whenever you manage to the best of your ability, despite all of the challenges and frustrations, to clear your mind long enough to perceive the magnificence of the biosphere, you are letting a spark reside within you. No matter what your station in life, regardless of your scientific expertise, whether you are new to environmental studies or a grizzled veteran, whenever you share your insights, heighten your aspirations, and provide a glimpse of these sparks to your family, community, students, clients, and colleagues, then you are confronting, in the most vital way possible, the perceptual challenge of global environmental change. It's in the act of gathering sparks that you are bringing the biosphere home.

2   The Experience of Globality

A Global Wave

I spent December 31, 1999 observing the weather and landscape, punctuated by spells surfing the Web and watching the news. It was a glorious winter day—seasonably cool, clear sky, and light winds. There was no snow on the ground, so you could wander unimpeded through the winter woods. After returning from a morning walk, I logged on to the Internet, curious to check on the impact of the Y2K bug in New Zealand and Australia. Throughout the course of the day I would recheck—how were things in Japan, China, Russia?

Like many, I was swayed by the media hype surrounding year 2000 matters, despite my judicious predisposition to view this New Year transition as one day just like any other. I prefer to herald and reflect on times come and gone during the Jewish High Holy Days when I can do so with structure and text to guide me. Or to trace arrivals and departures with changing seasons and the observation of cycles in nature. After all, what's so special about a number?

Yet 2000 carried an unshakable mystique. For how long had that number stood before us, beckoning the future, a guidepost to mark human progress and survival? And for how long did it seem unattainable, distant, a murky dream? It stood out, too, as the year I would turn fifty, and when the postwar baby boomers would all ponder the new century from the depths of middle age and their irretrievable youth. No matter how much I would feign indifference, the millennial passage gave me much to think about, harboring a potent brew of celebration and trepidation.

On television, CNN had cameras stationed in an array of global cities reporting on millennial events. It gave me great pleasure to see smiling, celebratory faces in these distant locales, people of diverse ethnicities, waving at the camera, signaling to millions of global viewers that all was well. For just a few moments, people forgot their differences, observing and enjoying something they all had in common—the passage of time on a revolving planet.

These waving hands, sprouting synchronously across time zones, reminded me of the "wave" at a ballpark in which fans rise spontaneously when it's their turn, simulating a swaying movement. I watched the year 2000 make its inexorable march across the globe, with people popping up when the shadow of midnight traversed their place—a global, group mug timed for the cameras, brought to your living room. Soon it would be my time to rise. Sentimental and naive perhaps, but I was overcome with emotion. This was an inspiring global connection. As I spent the day meandering from the New Hampshire woods to the media circuitry and back again, the concept of global interconnectedness passed from the abstract to the tangible. Yes, these faces in disparate places, across this inconceivably large planet, share a similar destiny.

On New Year's morning, the *New York Times* had a sparse and bold banner headline—"1/1/00." Pages A2 and A3 contained a series of ten symmetrically aligned, vertical stories, covering the new year in places near and far, from Timbuktu to Moscow. Across the top of the page was the compelling observation—"As the Dark Sweeps the Earth (Well, Most of It) . . . the Peoples Sing and Shop (and Still Wage War)." Much of the newspaper was devoted to coverage of the millennial passage. Flipping through the pages, I participated in a recursive spectacle—humans observing each other observing each other. "Twirling Globe Stops to Greet 2000, One Midnight after Another."

In stark contrast another headline beckoned me—"Future Threatens a Place without Calendars." A *New York Times* correspondent reports from Ekambu, Namibia, describing a "tiny village of mud huts and wildflowers, in the craggy, arid hills" where "there are no calendars, no electricity, and no words in the native language for millennium, computer, or Y2K." One of its residents, Maverihepisa Koruhama, tells the reporter, "when the thunderstorms start and the leaves grow from the ground, that's how we know it's the new year." Much of the article cites the various challenges to this way of life, not all of which are welcome, and how the thrust of development will inevitably catapult even this remote Namibian village into the global economy. Soon, the article implies, Koruhama's children will rise up and wave when the sweep of midnight passes through their village.[1]

Global awareness is a cognitive faculty linked to one's cultural situation. The globe appears very different from the hill country of New Hampshire than it does from downtown Los Angeles. Astronauts viewing the earth from space have a stunningly different perspective than Koruhama does from remote Namibia.[2] Yet anyone who was tuned to the television set on the night of December 31, 1999 had a similar experience. The global wave was a virtual sensation, presenting millions of viewers with the image of countless people connecting with each other—waving, nameless figures on a global television screen.

The image and metaphor of the global wave stayed with me for some time. It spurred me to think about all of the other ways that I observe the patterns and processes of an interconnected globe—icons of Earth from space, threats of nuclear annihilation, worldwide electronic communications, and endless talk about the effects of a global economy. In just half a century, during the span of my life, these reminders have become almost commonplace. I take it for granted that I watch a newscast that is generated in dozens of locations around the world, wear a shirt made in Malaysia, and work with a student who commutes from the Philippines. The global wave on New Year's Eve was a metaphor for the ocean of changes we are all exposed to as the global economy transforms the world landscape.

Yet I'm not sure whether the conceptual proximity of these daily reminders contributes to a greater sense of global environmental awareness. Here is the perceptual challenge that is the subject of this chapter—how might these reminders of global interconnection serve to promote greater awareness of global environmental change? Or to ask the same question metaphorically—where's the biosphere in the global wave? As an environmental educator, my aspiration is to use the daily reminders of global interconnection as learning opportunities, openings from which to view the biosphere.

The concept of global interconnection is particularly challenging because it appears in three intertwined venues. In the routines of making a living you are immersed in a global political economy. As a living organism, you participate in the ecological cycles of the biosphere. As a human being who wishes to attach meaning and significance to your actions, you are engaged in the myths and stories of a global stage. The strategy of this chapter is to reflect on these venues as a means of perceiving global environmental change. As I do with my students, I ask you to approach this material autobiographically, and to think about how you've come to learn about global interconnection. I describe this process as the "experience of globality." Here's how we'll approach it.

In the first section I show how one's experience of globality entails a series of conceptual leaps which require metaphor, ideology, and cognition, what historian Benedict Anderson describes as "imaginary linkages." Forms of mass communication, such as the newspaper, television, and the Internet catalyze these linkages. The breadth of your global experience also influences them. In the second section, I describe the intellectual and historical emergence of the idea of global environmental change, and show how it's grounded both in one's personal experience and an expanded scientific view. Third, I consider how the intertwined venues of globality, as represented by the global marketplace and the biosphere, yield rich metaphors. The interpretation of those metaphors provides a means for learning to perceive global environmental change.

Imaginary Linkages

Like millions of modern citizens, I was prepared for the millennial passage by virtue of our common morning habit—reading the newspaper. When I was five years old my desire to read was catalyzed when I noticed the extensive columns of numbers on the sports page. I came to realize that each morning's box scores and standings had rich stories to tell, the results of games won and lost in cities around North America. Brooklyn 8, St. Louis 3! Within a two-inch box was a record of the game, with enough information for me to create a plot—who played well, what the weather was like, when the fans cheered. It amazed me to think that all of these games were happening at the same time. Every day the plots of dozens of baseball games unfolded, linked together by numbers and text.

As I got older I discovered new corners and sections of the newspaper. On the weather page I learned that different regions of the country had diverse climates. How could it be that it was ninety degrees in places like Miami and Phoenix when it was twenty degrees in New York City? Eventually I discovered the front page itself, where stories of the entire world appeared. I wondered about these people and places—Khrushev, Eisenhower, the Berlin Wall, the Suez Canal. A twirling globe, for sure.

To think that so much could be happening at one time and it could all be described every day in the pages of a newspaper—extraordinary events, commonplace situations, and tales of terrible things. The newspaper served as scripture. Much of my experience of the world was (and still is) formed by the hours I've spent perusing the newspaper.

Today my reading habits have changed. I often scan the news online and reserve print newspapers for Sunday. As an educated adult I'm more aware that I should view what I read with detachment and skepticism. Yet the daily news, whether online or print, still serves a vital connecting function. I bear witness to the simultaneity of thousands of emerging events. Just knowing about these events reassures me of the validity of my own experiences and stories, as if whatever I do too is covered by some omniscient newspaper.

Benedict Anderson, in his insightful book on the origins of nationalism, *Imagined Communities,* suggests that the newspaper is a conceptual bridge allowing people to conceive of a vast expanse of disparate, remote events contained within a larger, cohesive framework. The idea of the nation requires an imaginative capacity because it links together so many diverse circumstances beyond what you would ordinarily encounter. The literary convention of the newspaper stimulates these imaginary linkages.

Anderson suggests you look at the front page of a newspaper and note the range of stories—an election result, a riot in a distant capitol, the discovery of a new planet, a scandal in a local bureaucracy. What connects these independent events? The next day's headlines might bring an entirely different arrangement of stories. "The arbitrariness of their juxtaposition or inclusion . . . shows that the linkage between them is imagined."[3]

This "imagined linkage" has two sources. The first, "calendrical coincidence," or the "steady onward clocking of homogeneous, empty time," provides continuity. If any story disappears for a while, readers don't assume that the place or event no longer exists. Rather, they understand the implicit reappearance of the event at some other time, similar to a character reappearing in a story plot. Second, the newspaper appears as a form of mass-consumption book, a "one-day best seller." Anderson describes this simultaneous imagining of newspaper stories as an "extraordinary mass ceremony."

The significance of this mass ceremony—Hegel observed that newspapers serve modern man as a substitute for morning prayers—is paradoxical. It is performed in silent privacy, in the lair of the skull. Yet each communicant is well aware that the ceremony he performs is being replicated simultaneously by thousands (or millions) of others of whose existence he is confident, yet of whose identity he has not the slightest notion. Furthermore, this ceremony is incessantly repeated at daily or half-daily intervals throughout the calendar. What more vivid figure for the secular, historically clocked, imagined community can be envisioned? At the same time, the newspaper reader, observing exact replicas of his own paper being consumed by his subway, barbershop or residential neighbors, is continually reassured that the imagined world is rooted in everyday life.[4]

Anderson is interested in how a complex array of simultaneous, diverse, and arbitrary events can be unified into a coherent, integrated framework. He suggests that the newspaper serves such an integrating purpose by offering a coordinated plot, which is shared by all of its readers. Although newspaper readers may interpret the various story lines differently, they are all working within a similar framework. They are linked together by virtue of their simultaneous reading. This is what he means by an "imagined community." The expression "imaginary linkage" suggests that each observer must form a mental image in order to make these connections real.

Anderson's work is useful not only for showing how the newspaper is a stepping stone for the idea of the nation but by extension how television and the Internet contribute to an idea of the globe. If you log on to the Internet each morning, you connect to an imaginary community, assured of its viability by virtue of its simultaneity, ubiquity, and reliability, and by knowing that there are millions of other people who are online, too. As Internet participation increases, as more of your daily correspondence appears there, as you visit more Web locations, dozens of interweaving plots unfold. You develop a set of imaginary linkages or images of an interconnected globe. This is an even more intricate, enveloping, and integrated international mass ceremony.

Interestingly, from the beginning, Internet icons have embraced whole earth imagery. Peruse any popular computer or science magazine and you'll find that many of the Internet-related advertisements convey images of branching networks, interpersonal connections, or human/computer interfaces superimposed on sundry globes and earth photos. The rhetoric of the global economy is often illustrated with proclamations of earth-scale influence and investment prowess. Investors seek global growth funds and equities. Those who want to handle your money assure its span and reach by displaying earth icons in their promotional literature.

When you reflect on the texture of your everyday experience, whether it's reading the newspaper, or logging onto the Internet, you can see how ideas such as nation and globe require an "imagined linkage." If you watched the global wave of the millennial passage, you imagined yourself as a participant in an enormous story with infinite subplots, all somehow connected to your daily experience. The context (the passing of the year) was of enormous significance yet utterly mundane. Its significance reflected the conveyed attention, turning midnight into a media spectacle. It's mundane in that it happens every day. The earth always turns from night to day. Only the numbers on the calendar change. You enter the global story as you wish or as your attention is so directed.

Think about all of the compelling events that you may have witnessed in conjunction with a global audience—the live explosion of the *Challenger* spacecraft, *Skylab* falling down to Earth, the nuclear accident at Chernobyl, the taking of hostages in Iran, the soccer World Cup. What makes them global is not just the fact that their outcomes are highly relevant to people around the world, but that you can imagine a global audience witnessing the same events—people whom you will never know in distant corners of the world are privy to similar stories.

These stories, whether you follow them in the newspaper, on television, or via the Internet, are no longer disparate or arbitrary. The very attention placed on them makes them seem as if they are of compelling importance and interest. You begin to follow stories that wouldn't have any impact on your life if it weren't for their media prominence. You are constantly observing the world as it observes you. And you'd rather not miss any of it.

It's important not to glorify the transformational impact of this simultaneous global imagination. On the morning of January 1, 2000, people weren't so inspired that they petitioned the turning of swords into plowshares. Or that they gained a new appreciation for the ecological fate of the earth. Or they suddenly became committed to the virtues of international cooperation and understanding. Nor is there any evidence in the trappings of the media global village to indicate that such sentiments are forthcoming.

However, it takes only a modicum of reflection to observe that distant events may well influence what happens to you. I vividly recall, and I'm sure if you are old enough you can too, the events leading up to and following the Cuban missile crisis. I lived near Kennedy airport (then called Idlewild) in New York City. We constantly listened to jets streaking over our house as they took off or landed. I remember my seventh-grade classroom during the crisis and how every time a jet passed overhead, a shriek would sound from the kids in the class, all fearful that war had broken loose. From that moment on, whenever I heard sirens in the middle of the night, I would turn on the radio to make sure it was just a local fire and not a nuclear holocaust. I would hear the news announcer on WCBS radio read the headlines and I was relieved that all was well. No nuclear war today! The radio served to connect me to the global community, assuring me of its presence. All of the imaginary linkages were in place—the baseball scores, the weather forecasts, the top ten singles. To this day, the media still allows me to conjure the viability of one integrated globe. But I learned then that a war breaking out in some distant corner of the world could have very serious consequences. There is no more profound way to think about global interconnectedness than to understand that a nuclear missile can cross the ocean in about half an hour.

What I wish to convey by explaining the concept of imaginary linkage is that it takes a chain of conceptual leaps and assumptions to perceive that an enormous globe filled with six billion people and several hundred countries has a shared destiny, a coordinated plot. What's most important is how you choose to interpret that linkage, the larger framework in which it fits. Often it spins like a global soap opera, a daily tale of mammoth proportions that you can tune into for your entertainment. In so doing, the experience of globality is no more than a narcissistic spectacle and you will see only your own reflection. But this remarkable conceptual chain is also a magnificent learning opportunity, a way to interpret the meaning of these events and to achieve deeper global awareness. If the daily news is literally a substitute for morning prayers, then your reading of the day should reflect on questions of meaning and value.

The conceptual proximity of globality, its metaphors, signs, and linkages, is an unprecedented opportunity to learn about global environmental change—its evidence is laid right before you. In the final section of the chapter, I discuss how one might use imaginary linkages as a method for interpreting the biosphere. Before doing so, I provide a historical context for the experience of globality by sketching a brief portrait of the intellectual origins of global change science. There is a correspondence between the expansion of global political economy and the idea of global environmental change. Within the crux of their co-emergence lies one's experience of globality.

The Roots of Global Change Science

Gradually and inevitably, with the expansion of world trade and its consolidation in a global economy, the development of air travel and rocketry, and the diffusion of electronic communications, the experience of globality becomes commonplace. Imaginary linkages are required to sustain its perceptual viability and various ideological systems serve to explain its rules and operation. For example, much of the popular rhetoric of globality is couched in the metaphors and concepts of global markets. Yet the patterns of globality also yield rich new ideas in environmental thought. Just as the idea of a global market conveys the worldwide movement of capital and commodities, so it also implies the flow of energy and matter through diverse ecosystems. Just as you learn to recognize how political events in the Middle East may have a dramatic impact on economic life in North America (the flow of oil), so you can perceive how the burning of oil fields can have long-term pollution effects, contribute to global warming, and threaten biodiversity.

Historically, the science of global environmental change is inextricable from an emerging awareness of globality. In the nineteenth century, voyages of discovery, often sponsored in search of new markets, also served to expand the global natural history database. As these expeditions reached their navigational extremes, the naturalist's catalog of landforms, flora and fauna, habitats, climates, and indigenous cultures dramatically expanded. By the late nineteenth century, much of the world's surface had been explored, connected, and colonized. Imperialists and entrepreneurs contemplated a global marketplace. Scientists were exposed to a broad range of geographic, geological, and ecological situations. The stage was set for considering the idea of an integrated global system, whether it was to establish an international protocol for politics and markets, to speculate on the movements of oceans and continents, or to theorize about the evolution of life on earth. Also, with this emerging cognition of globality, for the first time, ecologically minded observers could contemplate the worldwide impact of humans on the landscape.

Global environmental change science has its origins in four roots, three of which flourished in the late nineteenth century, parallel to the settlement of the globe by European expansion. First, the recognition and subsequent investigation of pollution, resource exploitation, and environmental degradation, spurred by a growing understanding of the prevalence and extent of these problems, yielded the science of conservation. Most global environmental science texts salute the extraordinary insight and prescience of George Perkins Marsh, a nineteenth century physical geographer. His classic text, *Man and Nature (Or, Physical Geography as Modified by Human Action)*, is the first systematic discussion of human ecological impact (1864), although terms such as ecology and evolution are never even mentioned. Indeed, Marsh's work was published only five years after Darwin's *Origin of the Species.* Marsh understood (influenced by the deforestation of New England and Europe) the extraordinary impact that humans have on the environment, and the global consequences of this impact. Marsh describes the purpose of his work "is to indicate the character and, approximately, the extent of the changes produced by human action in the physical condition of the globe we inhabit; to point out the dangers of imprudence and the necessity of caution in all operations which, on a large scale, interfere with the spontaneous arrangements of the organic or inorganic world; to suggest the possibility and the importance of the restoration of disturbed harmonies and the material improvement of waste and exhausted regions."[5]

Second, the growing sophistication of geology, biology, and chemistry, and the increasing interest in natural history, informed by scientific world voyages such as those of Humboldt, Wallace, and Darwin, among many others, catalyzed the expansion of a global natural history data collection process. This geographic breadth provided the natural history and geological experience with which to consider evolutionary explanation. The further people voyaged around the globe in search of natural history data, the more they were able to travel conceptually in both space and time. These voyages, coupled with a broader scientific understanding of natural systems, contributed to the first conceptualizations and use of the terms *evolution* (1859) and *ecology* (1866). Peter J. Bowler, in *The Norton History of the Environmental Sciences,* notes that "one of the most important developments within the cultural framework of modern science is the emergence of this awareness that Nature has a history that determines its present structure."[6] The science of global environmental change requires broad awareness of spatial and temporal scale.

Third, with the expansion of European influence and power, entrepreneurs and imperialists considered the long-term viability of global markets. Some of their agents wondered what the consequences of global expansion would mean for human populations or what later would be called carrying capacity—the relationship between world resources and human well-being. In 1891, E. G. Ravenstein is among the first to ask the poignant question, How many people can the earth support? His paper, "Lands of the Globe Still Available for European Settlement," offered a systematic analysis of the relationship between natural resources, technology, and population on a global scale.[7]

A fourth and crucial branch of global environmental change science emerged in the 1920s with the formulation of the biosphere concept. This idea required the integration of evolutionary and ecological ideas with an understanding of biogeochemical cycles as well as a broad geological perspective. In 1926, Vladimir Vernadsky wrote an entire book on the biosphere in which he essentially laid out the full ramifications of the co-evolution of life and the physical environment. In the introduction to the 1998 (first English translation!) edition of this seminal work, *The Biosphere,* an international consortium of global change scientists describes its significance. They cite three empirical generalizations that exemplify Vernadsky's concept of the biosphere. He observed that life occurs on a spherical planet and was the first person in history to come to grips "with the real implications of the fact that Earth is a self-contained sphere." He understood that "life makes geology" and that "virtually all geological features at the Earth's surface are bio-influenced." And he theorized the evolutionary implications of these processes, showing how the "planetary influence of living matter becomes more extensive with time." Increasingly, as the "spectrum of chemical reactions engendered by living matter" expands, "more parts of Earth are incorporated into the biosphere."[8]

What's particularly remarkable about Vernadsky's work is how he could formulate this biospheric perspective without access to satellite photographs or any of the advanced electronic instrumentation that is now taken for granted. He was the first to conceive of ecology and evolution as planetary sciences. Dorion Sagan, who along with Lynn Margulis, has been instrumental in resurrecting the importance of Vernadsky's work for contemporary global change science, provides an interesting perspective: "In a sense, Vernadsky did for biological space what Darwin did for biological time: he showed that the main traits, the scientific character of life as a whole, could be best grasped on a global scale, one that encompasses space and radiations from the sun. As Darwin led life to contemplate its ancient past by showing how all living beings descend from the same ancestors, Vernadsky extended the province of inquiry from local investigation to life on a planetary scale."[9]

What's impressive about the intellectual emergence of all four roots of global change science—conservation, ecology and evolution, carrying capacity, and biosphere studies—is how its founders develop such extraordinary insights by integrating their local observations with broader spatial and temporal considerations as influenced by their global travels. Ravenstein's carrying capacity calculations were abstract and primarily mathematical, but Marsh, Darwin (and his contemporaries), and Vernadsky all developed their insights based on their fine naturalist skills. What perceptual faculties enabled them to conceive this emerging biospheric perspective? A close reading of Darwin, Marsh, and Vernadsky reveals their interest and skill at perceiving local natural history by cultivating both sensory awareness and studying the science of natural history. I return to this idea in more depth in chapters 4 and 5.

I conceive of this period, roughly from 1860 to 1930 *as the roots of global change science.* Parallel with specific historical developments—the culmination of European expansion, the first phases of economic globalization, the internationalization of war, and the spread of high speed transportation—are the first conceptualizations of ideas such as evolution, ecology, biogeochemistry, carrying capacity, continental drift, and the biosphere. No causality is implied here. Rather I wish to indicate how the experience of globality involves two braided conceptual streams.

Apocalypse and Affluence

With the development of nuclear weaponry and long-distance rocketry, influenced in great measure by the Cold War competition between the United States and the Soviet Union, a new phase of global awareness emerged—*apocalypse and affluence.* As an elementary school student in the late 1950s, my experience of globality was colored by two distinct and contrasting impressions. My memories of this time lurk mysteriously, both vivid and dreamlike, as shadows of my youth. They are the light and dark sides of my childhood imaginary linkages to a nascent global perspective.

I understood the phantom of unthinkable war, and the mysterious but chillingly real prospect of instant death from a thermonuclear exchange. Why else would we practice air raid drills? Yet I also remember the glowing promises of an unbounded affluent future—the prospect of extraordinary technical advances and universal wealth, a globe united by science. I poured through the pages of a *Life* magazine that gave detailed accounts of the International Geophysical Year, proclaiming the cooperation of international scientists. I recall the principal getting on the loud-speaker in sixth grade to tell us about the successful first rocket trip into outer space. It was an era of scientific optimism under a shadow of cold, dark fears.

In retrospect, I can summon three stunning images of the early 1960s, all of which stirred my imagination, and were important formative influences in how I learned to conceive of the planet. Between 1962 and 1964, in my early adolescence, at a time when developmentally I was just about ready to think about bigger pictures, I experienced the Cuban missile crisis, the New York World's Fair, and the publication of Rachel Carson's *Silent Spring*. My favorite exhibit at the World's Fair was the General Motors city of the future. Here was an affluent, streamlined world, where there were no hints of disease or poverty, promising the very best of technical know-how. Outside the fantasy of the World's Fair, these images were sandwiched between the horrifying notions of stockpiled nuclear missiles and a dying, poisoned planet. Anyone who grew up in this era was exposed to scientific and popular representations of the future that skipped confusingly between images of apocalypse and affluence.

While the world's military industrial complexes were counting missiles, many geographers, ecologists, and conservation-oriented scientists were studying the impact of humans on the earth's surface. These studies, which received scant public attention at the time, carried a different foreboding, the Malthusian specter of population overrunning resources, proclaiming a very different vision than the optimistic, green revolution projections of plentiful food and filled tummies. In 1954, Harrison Brown wrote *The Challenge of Man's Future*, an ecological overview of world resource distribution. In 1956, an unprecedented symposium convened to discuss the ecological impact of human settlement, culminating in the classic work edited by W. L. Thomas, *Man's Role in Changing the Face of the Earth*. Motivated in part by the "one world" shadow of nuclear weaponry, these studies were undertaken with the whole globe in mind, developing methodologies and interpretive approaches that laid the groundwork for an entire generation of conservation-oriented global change studies.

The Emergence of Environmental Studies

About ten years later, in conjunction with the intense social activism of the late 1960s, Earth Day, and the blossoming of environmentalism, there was a growing awareness of global ecological issues. As a student at New York University I had a weekly routine in which I would visit the 8th Street Bookstore in Greenwich Village to check out the latest titles. One evening I noticed an unusual oversized book with a big picture of the earth from space on the cover—*The Whole Earth Catalog*. I grabbed it off the shelf, bought it, and returned to my apartment, reading it as I walked through the streets and rode the subway home. My head was buried in this book for weeks. It not only provided its readers with a fine compilation of resources but it offered a conceptual path for thinking about them—a unique blend of systems thinking, ecology, community, and appropriate technology, all revolving around a vision of the "whole earth." This global, ecological vision inspired me for years, providing a framework for learning and thinking, eventually leading to a career in environmental studies. *The Whole Earth Catalog*, with its picture of the earth from space on the cover, reflected the development of a new environmental perspective, concerned with both local community issues, and broader global questions. Over a two-decade stretch, roughly between 1968 and 1990, came a third phase of global awareness, *the emergence of environmental studies*.

As a means of considering the growth and development of contemporary environmental thought, I ask my students to take ten minutes to come up with as many words or expressions they can think of with eco-, green, or environmental in the prefix. A group of fifteen students will typically list well over two hundred combinations—from ecopsychology to green business. What's particularly impressive about these lists is that almost the entire lexicon emerged after 1968. The influence of environmentalism, integrated with the widespread media distribution of earth imagery and satellite data, the development of an accessible international communications network, and growing awareness of the scale of pollution and habitat degradation, together laid the groundwork for an ecological perspective on global matters. During this twenty-year period, thousands of colleges and universities initiated environmental studies programs and courses, environmental education became an international network of teaching ideas and curriculum, and the literature of environmental studies proliferated.

Examples of this impressive growth of global environmental awareness abound—the Club of Rome limits-to-growth studies (and their sequels) that used computer models and systems thinking to assess the future of world resource use,[10] the founding of the Worldwatch Institute and its research program focused on the environmental impacts of global economic growth,[11] the publication of popular books such as the

Scientific American special issue on the biosphere,[12] the growth of an international network of environmental scientists and policymakers who attempted to develop ideas, protocols, and research agendas from an ecological, planetary perspective. In one generation, an expansive and comprehensive environmental literature emerged, spreading throughout the sciences and humanities, inspired by a growing realization of "whole earth" priorities. This twenty-year conceptual splurge reflected an environmental response to the experience of globality.

Twenty-First Century Global Change Science

By the late 1980s, another phase of global awareness emerged, the *origins of twenty-first century global change science.* This was catalyzed by the phenomenal expansion of data collection resulting from the combination of computer speed, the Internet, and satellite imagery—enhancing the senses scientifically. Consider the impact of the following trends: an international system for the gathering and dissemination of ecological data, the emergence of geographic information systems, dynamic computer modeling of atmospheric systems, the ability to measure energy fluxes and biogeochemical flows, the development of paleoecological dating techniques. These technological achievements, along with dozens of other advances in computerized, electron microscopic, and telescopic research, expanded the conceptual reach of thousands of scientists, allowing for new ideas in the spatial and temporal dimensions of environmental information.

This prolific collection of global environmental data contributed directly to the idea of global environmental change, a term that came into widespread use in the late 1980s. Dozens of international, interdisciplinary consortia convened to collate this information and place it in theoretical perspective. Particularly influential was the massive anthology *The Earth as Transformed by Human Action,* which surveyed global and regional environmental changes from a biospheric perspective. This spawned the publication of a new generation of global change textbooks.[13]

Also in the late 1980s, as more data were collected from diverse earth habitats, conservation biologists and ecologists hypothesized the prospect of a forthcoming mass extinction, inspiring the genesis of biodiversity studies.[14] With precise carbon dioxide measurements and better techniques for tracing the flow of atmospheric energy, climatologists developed an understanding of global climate change.[15] The combination of atmospheric science, biogeochemistry, evolutionary thought, and microbiology contributed to the Gaia hypothesis, a spectacular theory regarding the coevolution of life and matter.[16] These conceptualizations, all of which require fluid interdisciplinary research, and an expansive understanding of the spatial and temporal dimensions of environmental change, contributed to the revitalization of the idea of the biosphere.

Global change science conceives of the earth as an integrated system of living matter with global physiological metabolisms, complex circulatory systems, drifting oceans and continents, and a "breathing" biosphere. These ideas, although rooted in a century of expanding scientific awareness, are still inchoate formulations. Compelled both by the rapid rate of many global environmental changes, such as global warming and species extinction, and the great difficulties in assessing the large number of complex variables involved in these issues, the field of global change science is accelerating without a comprehensive theory of environmental change. Although there have been many intriguing conceptual breakthroughs, using relationships like scale, magnitude, flow, and trajectory, it seems that global change science is in a period of paradigmatic flux, and its own practitioners can barely keep up with the patterns and trends.[17]

This remarkable conceptual pace yields ripe metaphors, as scientists and their interpreters employ imagination to speculate on the full richness of this empirical work. The idea of the biosphere emerges as the most resilient global change metaphor of all, as it encompasses the dynamic, swirling motion of the ever-changing relationship between life and the planet. A survey of some of the more popular recent books on the biosphere suggests the necessity of metaphor as interpretive tool—Lynn Margulis writes about a *Symbiotic Planet*, Tyler Volk describes *Gaia's Body*, it's "biochemical guilds," and the "global breath of life."[18] In chapter 5 I discuss the use of imagination in perceiving the biosphere. For now, I wish to indicate that even sophisticated global change scientists require imaginary linkages to support their empirical work.[19]

Historically, the experience of globality emerges through both the expansion of global political economy and a deepening awareness of global ecological interconnectedness. These themes can be described both theoretically and metaphorically as globalization and the biosphere. Both themes have growing bodies of literature which purport to explain the economic and ecological patterns of an integrated globe. And both require imaginary linkages to be sustained. A global marketplace embodies an integrating framework that consists of stories and images of worldwide business and shopping. Understanding the bio-sphere entails imagining such concepts as biogeochemical cycles, biodi-versity, and weather systems.

Finding the Biosphere in the Global Economy

Nevertheless, global images and metaphors are more typically inter-preted as narratives of global economy. To validate this, consider how the experience of globality pervades your life. In the course of your rou-tine activities, what encounters remind you of your relationship to an in-terconnected globe? Do they remind you of the global economy or the biosphere? For most people, these encounters are embedded in the ma-terial reality of everyday events. With just a little bit of extra probing and interpretation, one's immersion in the global economy can be the source of biospheric narratives too.

Here are some plausible scenes. You wake up each morning, drink or-ange juice from Florida, put on clothes made in Asia, drive in a car made in four different factories on three continents, go to work for an interna-tionally owned conglomerate, perform international business transac-tions on the Internet, and take vacations abroad. Whether you are an unskilled laborer, or a highly trained professional, chances are your oc-cupational prospects are linked to the flow of international capital, and you probably move around a lot. Speak to a North American urban taxi driver and you are likely to find a recent immigrant from Asia, eastern Europe, or Latin America, who divides the year between his new resi-dence and the old country, earning money in one place to deliver it else-where to those family members who cannot possibly generate the same income at home. Or speak to a corporate executive who spends hun-dreds of hours attending meetings in various international capitals of fi-nance and finally returns home to an affluent suburban dormitory. Earning a living, consuming a meal, buying clothes, participating in popular culture—these are the overriding, material dynamics of globality.[20]

These experiences require imaginary linkages. The global economy is a myth of sorts, made coherent by the assumptions of unlimited eco-nomic growth, the imperial reach of unbridled transnational capital, and the broadcast of global popular culture. On a Saturday evening in the middle of December there are hundreds of thousands of cars and shop-pers moving in and out of elaborate indoor shopping malls, fully partic-ipating in the intricate mesh of the global sweep of commodities,

without questioning the integrity or viability of the story they are enact-
ing. There is a subtle reassurance in knowing that the goods are on the
shelves, that there are millions of people all over the world who are
shopping at this same moment.

This is merely a veneer of globality. Whether it's through their employment or their shopping habits, most people are aware of their active
participation in the international marketplace. But I don't imagine that
the average mall shopper considers the flow of matter and energy represented in his or her purchase, or how bottled water is a product of the hydrological cycle, or the long-range ecological impact of those purchases.
These processes are rarely considered. They are completely hidden from
view. It takes several degrees of abstraction and a disciplined intellectual, if not moral, commitment to find the biosphere when you're shopping. There is little correspondence between one's experience of
globality in the shopping mall and the idea of global environmental
change.

Yet if you pay even remote attention to the popular media, you will inevitably be exposed to images of the biosphere and messages about
global environmental change. The Weather Channel's great appeal may
be attributed in part to its prolific satellite photographs and to its coverage of weather disasters. The appearance of the West Nile virus in New
York City spreads fear and uncertainty. Whether it's the burning of tropical rainforests, the effects of El Niño, or reports of endangered species,
you are likely to encounter some daily story or image that reflects global
environmental change. My experience suggests that these images are extremely provocative and intrinsic to one's experience of globality. It's
just that it is not immediately apparent how to make sense of them.

Global environmental change is simultaneously ubiquitous and invisible. Without sophisticated scientific instruments, you can't actually see
the ozone hole, but you are told to keep out of the sun and limit your exposure to cancer-causing ultraviolet. Biodiversity is a theoretical concept reflecting the rich fabric of life that sustains any ecosystem. It takes
an experienced observer to assess various biodiversity indicators.
Global warming portends the alteration of world-wide weather systems, atmospheric and biogeochemical processes that pervade every
community, but one can never be sure if any single weather event can be
so attributed.

Like an uncommonly warm winter day, these issues sprout into
awareness, only to disappear as the temperature returns to normal. If
you observe landscapes and habitats, and wonder at all about ecosys-

tems and weather fronts, you're aware of the possibility of global environmental change. Unless you're directly engaged in a specific, scientific study, most of the changes that concern you occur at intangible, abstract levels and require the skills of a practiced observer. Who is actually witness to extinction? Who can observe the ecological impact of an invasive species? People who are familiar with environmental issues and concepts know that observing global environmental change is a hefty conceptual challenge.

A fine paradox emerges. Global environmental change is too elusive to grasp, yet too profound to ignore. Not yet the province of concentrated public attention, it appears more subtly, through its images and metaphors. Not easily understood, it leaves its marks and trails nevertheless, in the form of local signs and global reflections. International networks of commerce and communication may hide the ecological origins of your daily life, but they bring images of the planet to bear on your every move, whether it's the Netscape icon of a comet passing over the globe, or a Coca-Cola ad panning the world's cultures for coke drinkers.

Just as Netscape and Coca-Cola now permeate the global economy, so do images of global environmental change pervade the airwaves. Portraits of the planet—obscure, intangible, metaphorical, symbolic—contain meaning as vast as the cultures and individuals who witness them. Global climate change and biodiversity are much more than scientific theories. They envelop the very origins and unfolding of the biosphere, thereby carrying collective traditions of story, myth, and meaning. Such is the richness and resilience that surrounds how global environmental change is perceived.

The very idea that something is wrong with the planet, that it's changing in unprecedented and unforeseen ways, creates a feeling of unease. You may not necessarily fully understand the scientific aspects of global warming, or even be convinced of its importance as an issue, but surely its prospect leaves some kind of impression. You may not directly witness the decline of an endangered species, but the image of widespread extinctions may raise feelings of loss. In chapter 3, I discuss the spiritual implications of such feelings in more detail, suggesting that global environmental change should be pondered from both ecological and existential perspectives.

The images of global change provoke intangible forebodings—the possibility that forces beyond your control may be at work, but you don't fully understand what they are or how they may impact you. In the absence of scientific assessment or practiced ecological observation, you are left with anecdotes and metaphors. My impression is that this is precisely how most people learn about global environmental change,[21] through a series of images and feelings—imaginary linkages—developed through casual observations of nature, glimpses of a few headlines, and some resilient media icons. These observations, taken together, may leave a diffuse anxiety—feelings of fear, loss of control, unease, and uncertainty—without any real point of convergence.

Using Metaphor to Perceive the Biosphere

If these feelings and images are the means through which most people learn about global environmental change, how might they be probed? As a means of approaching this issue, I often ask students to engage in spontaneous creative writing using the imagery of ozone hole, biodiversity, global warming, or species extinction. What emerges is a series of interesting metaphorical relationships that exemplify the unusual imaginary linkages that surround global environmental change. The following passages are a compendium of these responses.

When I first learned of the ozone hole, I immediately thought of the emerging bald spot on the top of my head. I know it's there because other people tell me about it. If I have access to properly angled mirrors I can see it for myself. I don't worry about it much. I can't reverse the aging process. In my insecure moments, it does leave me feeling somewhat vulnerable, as if my head is no longer protected. Similarly there is a hole in the stratosphere, the protective covering of the earth. I can't see the hole but scientific instrumentation reveals its presence. I have no reason to doubt this.

When I stroll in the sunshine I can no longer be as carefree because dastardly, invisible ultraviolet rays may be causing trouble. It's disquieting to ponder a big hole in the sky, letting in barbarian rays, and wreaking havoc with earth's immunology. The ozone hole signifies a wound in the global skin—a protective layer, a casing—serving as a boundary between the warm, intimate, organic soup of the earth and the cold, invisible, cosmic rays of outer space. How vulnerable it feels to think that the skin may be open, and there is a small wound allowing the entry of the forbidden.

Did you ever leave the house to go on vacation, drive for about thirty minutes, and then wonder if you left the burner on? I do that all the time. It's a nagging feeling. You know that you're probably playing mind games with yourself, working through the transition from stress to relaxation and it's highly unlikely that the stove is really on. But what if it is? You continue on your trip refusing to yield to your paranoia. You call a neighbor who'll check for you. This is how I conceive of global warming, except it's more like industrial society left the stove burning, but everyone continues on their trip anyway, assuming the neighbor will turn down society's oven.

Global warming is a thermometer, the one you stick under your tongue to measure your body heat. It is the metabolic indicator of global physiology, the ultimate arbiter of the global energy budget. Industrialization is cooking the raw planet. As the world gets warmer the global metabolism changes accordingly. What tempests may be unleashed when the circulation spins just a little too fast. Are the fevers, chills, and floods—recent patterns of extreme weather—the harbingers of a wavering physiology? What is the connection between burning rainforests, fires in Florida, smog over Malaysia, and crushing Pacific coastal storms?

The idea of species extinction leaves a big pit in my stomach, a hollow feeling, the impression of something vanishing. It's a feeling of loss and sadness as if someone took a good friend away on an extended journey never to return, but I'm not exactly sure who was taken and why she left. This corresponds well with a popular biodiversity metaphor, the idea that every endangered species is like an encyclopedia of earth history and its extinction is the equivalent of burning a library. So many friends and books that you will never get to know.

We all become accustomed to vanished landscapes—the special place you explored as a child, the newly developed wetland, the farm and forest turned mall. Whenever I observe such development I wonder about the disrupted habitats and displaced species. I ponder homelessness. What happened to the flora and fauna that used to reside there? Will they migrate and relocate or gradually perish? I recall how the enclosures of eighteenth-century industrial Britain uprooted thousands of agricultural poor, sending them wandering to the cities. Habitats and cultures are formed over decades and centuries. They disappear in seconds and minutes. Is there a correlation between vanished landscapes, loss of biodiversity, and the global trail of refugees?

Such narratives evoke full meaning when they are supported by sound global environmental change science. This provides depth and contrast, enabling you to ground the metaphors, take them out of the present moment, and situate them in the broader reaches of earth history. For example, as you gain familiarity with ecology and evolutionary science, you learn there are "background" extinction rates, and the rise and decline of species ebbs and flows like the waves of human empires. One can't mourn every loss. Geology and climatology teach you that the earth is always changing temperature and sometimes in rather extreme ways. The metaphors promote meaning, helping you consider the implications of these trends, aiding your perceptual breadth. They bridge the unsettled ground of intangible forebodings and the purposeful realm of scientific exploration. They help you pose the question—what does global environmental change portend for the human condition? And how is the human condition a reflection of earth history and biospheric processes? Through these questions one's experience of globality is used to perceive biospheric processes.

Asking such questions is a way of keeping global change in mind. Exploring metaphors links these questions to your deepest ecological and spiritual concerns. Global warming, biodiversity, ozone depletion, and species extinction are not just words lifted from an environmental science text. They reflect the prevailing conditions of the biosphere. And as you are of the biosphere, they are a means of describing your condition too. The metaphor is a medium through which that condition may be explored. With deliberation and imagination, the metaphor removes intangible forebodings from the recesses of your daily experience and shunts them to center stage.

What does one imagine the globe to be? An interconnected mesh of Internet shopping, mall complexes, and theme parks, serviced by a global resource bank of minerals, crops, and machines? Or a biogeochemical circulatory system of coevolving life and matter, linked by networks of biosphere preserves, supported by a rich matrix of biodiversity and living landscape? The experience of globality demands interpretation and deliberation. Global interconnections are revealed in many different ways. The global wave is more than just a global, group portrait—it's the whirling motion of the biosphere as it permeates the planet. Educators and concerned citizens have a rich opportunity, finding ways to use the dozens of daily reminders of global interconnectedness as a means to interpret the biosphere.

In perceiving global environmental change, the biosphere becomes the integrating metaphor of global awareness. This requires powerful imaginary linkages. You have to make its story your own. Global change science provides the text for the story—the daily drama of worldwide weather, the movement of peoples and species, the travels of a Carbon atom—all read and interpreted as the news of the biosphere. Whole earth imagery and metaphors contribute fine illustrations—photos of the earth from space stimulate great imaginative depth. What is the purpose and meaning of the biospheric journey? You find the biosphere in the swirling circulatory systems of the spinning globe, the rich natural history of your home place, and the unfathomable dimensions of the human heart. As the world turns, infinite complexities of ecological and evolutionary plots unfold. Learning to perceive the biosphere, you always keep global change in mind.

The doors of biospheric perception are slightly ajar. The weather report is no longer just a way to figure out what to wear; it's a means of tracking atmospheric circulation around the globe. Your morning walk isn't just an aerobic workout; it's a way to observe the living landscape of flora and fauna. The study of global environmental change is more than encountering feelings of loss and vulnerability. It broadens and stretches your mind in inconceivable ways, making the entire biosphere your learning pathway, cultivating wonder and awe in the mystery of creation.

Here is another way of describing the perceptual challenge of global environmental change—learning how to reflect on the experience of globality from a biospheric perspective. Inevitably this raises profound questions of meaning and purpose. How does your awareness of the biosphere inform your daily life? How does it influence your view of creation and extinction, your sense of wonder, your human identity? What is the responsibility of humanity in regard to global environmental change? Every reminder of global interconnectedness, every observation of the biosphere is a bell ringer for such questions.

This is one of the most difficult obstacles to contemplating global environmental change. It raises deep, penetrating questions about the human condition in relationship to the biosphere that are almost impossible to answer. The more you learn about the biosphere, the more you cultivate awe for the robust ecological and evolutionary matrix of life—the wonders of creation. Yet you simultaneously observe its fragility—the prospects of mass extinction weigh heavily. Finding the biosphere entails both ecological and spiritual contemplation, and it is to that effort that we now turn.

3   Keeping Global Change in Mind

From a Peep to a Clamor

I live several hundred yards from an abandoned beaver pond. Several narrow streams plunge from the hills into a topographic bowl. Some time ago, beavers dammed the main outlet, plugging the outflow passage, turning the bowl into a wetland. Only two houses adjoin the wetland, and they are sufficiently set back so that the landscape seems hidden from view, a minisanctuary nestled in the rugged hills of southwest New Hampshire. Wildlife is active and prominent—great blue heron, wood duck, migrating warblers, fisher, porcupine, red fox—just about anything you might expect to find in the northern forest.

Sometime in mid-April, when the snow is melted, and the smell of spring is in the air, you can hear the first rhythmic clatter of spring peepers. As soon as the waters are free and open, frogs become active and begin their reproductive sequence. Their distinctive vocalizations, specific to the species, are mating calls. The peepers provide the first amphibian sounds of spring. Although they are most active at night, they begin their calls in the afternoon. Because there are so many vernal pools near my house, for much of April I am surrounded by this primal sound of spring.

On a pleasantly mild, clear evening, I ambled over to the edge of the pond, intent on listening to the peepers. You have to tread quietly. The volume of the peepers is directly proportional to how much noise you make. With your increasing stillness, the sound of the peepers grows in volume and intensity. Sitting silently is a gateway to what becomes a deafening roar—from a peep to a clamor.

After a few moments, you're immersed in the resonance, vibration, and deep mystery of this ancient, primordial sound. Frogs have been around since the Triassic period, so it isn't too far off to date them at

220 million years old. They have been here for a very long time. I like to chant with the peepers, singing low harmonic improvisational melodies under and between their beats, low enough to join their chorus, without really being heard. In so doing, I feel as if I'm listening to the ancient rhythm section of evolutionary time. The clamor of peepers resonates through my entire body until I feel as if I'm bathing in symphony—spring peepers surround sound.

Yet there is no mistaking the awkwardness—the sense of intrusion, or even trespass—that I am improvising with a band making music beyond my understanding of rhythm and melody. I am a sheer beginner, like a child exploring a toy piano, probing the musical spectrum for the first time. I wonder whether I belong here, if my singing is some kind of violation, if I haven't practiced enough. What can I know of 220-million-year-old songs? I experience a tentative exuberance, wavering between immersion and self-consciousness, tangled in a laboratory of sound and spirit.

I glance upward at the sky. It is a spectacular heaven, moonless, filled with a glistening stellar array. Each peeper's song corresponds to a star. By this measure I contemplate the inconceivable depths of space and time. Each star is a beacon of memory—frozen light—traveling across the cold, mysterious expanse of the limitless void. The Tibetan Buddhists suggest that there are a billion thought moments in each second and for that instant, as I sat by the beaver pond, I believed that to be true. But I couldn't sustain my focus. My eyes looked away from the heavens and my ears retreated from the peepers. Feeling I just couldn't take in any more (was I lazy, fearful, or prudent?) I returned home to the safety of the routine and familiar.

I hadn't read the day's news so I went to www.ennnews.com[1] to check out the environmental headlines. I noticed a lead story describing how a scientific study in Costa Rica is finding links between global climate change and the decline in several species of tree frogs. Apparently subtle shifts in rainfall are causing habitat changes which are disrupting the life cycles of these frogs.

I wondered whether the spring peepers in my ecological neighborhood were subjected to similar environmental challenges. Of course, I knew this to be true. Much has been written about the startling increase of deformed frogs in some Northeast and Midwest wetlands. Many causes are suspected, including increased ultraviolet exposure and the presence of endocrine disruptors, although there is no definitive statement on this. It occurred to me that a metaphor of my childhood was meeting a metaphor of midlife head on. Silent Spring, I'd like to introduce you to Global Warming.

I resume surfing the net and visit www.nytimes.com to read the latest from Kosovo. Images of displaced refugees and bombing runs flash across the screen—pixels of suffering eclipse my peeper-inspired reverie of just a few moments ago. I wonder about the pain and terror experienced by people who are forced to leave their homes at a moment's notice. What if I had to endure such hardships? I think about the ecological destruction levied by the bombs. What if the beaver pond across the road became a battleground? A penumbra of anxiety passes over me.

Here in the course of merely an hour I had relived one of the major spiritual themes of my life—experiences of wonder and inspiration coupled with moments of dread and apprehension. Today I live the "good life" (so many opportunities and experiences) for which I am deeply grateful, yet my world is filled with what Freud described as "chronic apprehensiveness,"[2] portents and forebodings—experiences of "existential anxiety"[3] in tandem with my sense of wonder. These are deep impressions that are at the core of my being in the world. Indeed, they are characteristic of our baby-boom, millennial, turn-of-the-century milieu—shadows and light, opportunity and foreboding, liberation and apocalypse, wealth and misery.[4]

I'm convinced that these tensions are intrinsic to contemplating global environmental change. Issues such as the loss of biodiversity, species extinction, and climate change are brimming with existential dilemmas. Learning about ecology and evolution, becoming familiar with the history of life on earth, attaining a biospheric perspective—these approaches to thinking expand the mind and enhance awareness, filling you with curiosity, wonder, awe, and praise. One is also struck by the sheer vastness of these earth processes—the expansiveness, grandeur, and complexity of the biosphere. When you perceive the human condition from this biospheric perspective, you contemplate both the crucial importance and utter insignificance of human action. What difference does any human action make when measured against the inconceivable breadth of space and time—4.5 billion years of earth history? Yet the inexorable march of humanity has brazenly transformed the face of the earth in a remarkably short period of time, yielding profound climatic and ecological changes. We are far less powerful than we think, but have much more impact than we can perceive.

To bring the biosphere home means to reconceptualize your everyday actions so they are informed not only by greater intimacy and familiarity with natural history, ecological relationships, and earth system processes—that is challenge enough—but also by questions of meaning and purpose. Teaching and learning about global environmental change covers the realms of both science and spirit. Learning to perceive the biosphere requires the deepest perceptual and experiential participation, the full involvement of your senses. Yet such immersion entails a sobering dose of humility—the impossibility of grasping the full measure of one's existence.

The chorus of the spring peepers is one of many nature stories that lead you to explore this interesting territory, field guides for keeping global change in mind. Their interpretation suggests a way to organize this chapter. First, I ponder the relationship between creation[5] and extinction as processes within the broad spectrum of biodiversity. Every species tells a story and each story contributes to the unfolding text of ecology and evolution. How might the life story and history of any species, as well as its prospects for survival, provide a means of assessing biodiversity and the human condition? By considering both the origins of species and their termination, and the role humanity plays in the process, you contemplate the ecological and existential aspects of biodiversity.

From this foundation, several tensions emerge. For whom do the peepers spring mating calls elicit wonder and curiosity? The cultivation of wonder has long been at the heart of environmental education. As such, it is also at the core of this book—the deep wish that wondering about the biosphere summons praise, compassion, and an ethic of care, the hope that there is still room for wonder in the hearts and minds of children and adults. But wonder is often met with indifference and the tension between the two is a profound educational and spiritual dilemma. What role does wonder play in perceiving global environmental change?

The provocative, inspiring peeper calls provide a song of hope. Their annual return is a reminder of the resilience of the biosphere, its depth and longevity, how it surpasses one's limited view and experience. Meanwhile, the global plight of amphibians at the hands of human development suggests the decline of the peepers, a time when their songs will no longer grace the countryside. Who can imagine a world without song? This tension of hope and despair emerges with virtually every global ecological issue. Wherever hope and despair reside, faith and doubt are in proximity. When faced with the prospect of amphibian extinctions, or any other complex global environmental situation, one's re-solve is continuously challenged.

Creation and extinction, wonder and indifference, hope and foreboding—these tensions, impulses, feelings, and considerations reveal the spiritual depth intrinsic to issues of global environmental change. It is hard to imagine one becoming fully immersed in the songs of the peepers without wandering through this reflective territory. Yet one's hesitation to consider these questions may serve to push the biosphere away, to deny its importance as a human concern. Learning to perceive the biosphere requires a special confluence of ecology, psychology, and spirituality. In contemplating the biosphere, you observe earth system processes through the lens of deep time, and in so doing, you inevitably reflect on the origins of life, the evolution of mind, and the purpose and meaning of human action.

Ultimately it is your deeds that count. Where does your encounter with the peepers leave you? If it engenders praise, in what way do you demonstrate your gratitude? What actions follow reverence, and how might they be considered and displayed? Keeping global change in mind requires ecological intimacy and spiritual concern. This chapter considers the resilient tensions that pervade such thinking, with special attention placed on the learning opportunities that may unfold.

Creation and Extinction

Consider, for a moment, the number of species that currently live on earth. If you are having trouble coming up with a number, relax because you're in good company. Between 1.5 and 1.8 million living species have been cataloged by taxonomists, but there are millions of uncataloged species as well, and some scientists estimate that there are as many as 30 million species living in the biosphere. This is an exceedingly high and extraordinary range considering the precision with which scientists can count most things. As Edward O. Wilson comments, "the number of species of organisms on earth is immense but still cannot be placed to the nearest order of magnitude," hence he calls the earth, "the unexplored biosphere."[6] This range can be attributed to both the large number of undiscovered species in relatively unexplored habitats (tropical forest canopies and deep ocean floors) and the fact that, according to Andrew Dobson, "the distribution of taxonomists is almost completely the opposite to the best current estimates of where the earth's biodiversity is found,"[7] that is, Latin America and sub-Saharan Africa. Of these 1.5 million known species, approximately 750,000 are insects, 41,000 are vertebrates, and 250,000 are plants. Remarkably mammals and birds constitute only 0.025 and 0.066 percent of the total respectively.[8]

Now consider this approximation (1.5-30 million) in relationship to the number of species that have ever lived on earth. Colin Tudge, in his survey of biodiversity *The Variety of Life*, addresses this:

So how many more species of all kinds of creatures might there have been over the past 3,500 million years, given that most are small and some have a generation time measurable in hours? It would be surprising if the total number of species in the past did not exceed the present inventory by at least 10,000 times.

In short, the number of species that have lived on earth since life first began could easily be about 400 million times 10,000, which is 4 million million or 4,000 (American) billions—roughly a thousand species for every year that life has existed on earth. Of course, these estimates may be out by an order of magnitude, or even by several orders of magnitude. But even if they were exaggerated a millionfold the total would still be vast; and far too great for any human mind to grasp.[9]

Throughout the 3.8 billion years of life on earth, there have been infinite combinations and permutations of genes, environments, and ecologies. The historical scope of species, living and dead, is incomprehensible. Only in the last one hundred fifty years have scientists gained an inkling as to the complexity of the evolutionary past, and as tracking techniques and fossil evidence are unearthed, more details become available. We are afforded only the barest glimpse into the multitude of ecological and evolutionary combinations. The emergence of earth systems sciences, including ecology and evolution, allow us merely to place a theoretical structure on this unbounded history.

If you wish to stay bounded, you can gain a temporary reprieve by retreating to a backyard, park, garden, or landscape where wild things grow. By virtue of sight alone, see how many different species you can identify. You'll be able to distinguish the larger species that don't move (trees, bushes, and wildflowers), and perhaps notice a few insects and small mammals. Depending on your natural history prowess, you'll be able to identify several dozen of these. But the minute you switch scale, you gain insight into the magnitude of species that reside at the edge of your awareness.[10]

Writing about the "unexplored biosphere," Edward O. Wilson contemplates the bacterial presence in common soil by offering a neat quiz: "Take a gram of ordinary soil, a pinch held between two fingers, and place it in the palm of your hand. You are holding a clump of quartz grains laced with decaying organic matter and free nutrients, and about ten billion bacteria. How many bacterial species are present?[11] Wilson suggests that no one really knows. *Bergey's Manual of Systematic Bacteriology* lists about 4,000 species, but recent research claims that the total number may range into the millions![12]

The number of species prevalent in any place is a function of habitat and scale. Species richness increases in tropical settings and declines toward the poles. The more microscopic your gaze, the more species you'll be able to observe. For example, the abundance of tree canopy and soil flora and fauna is virtually invisible and unexplored, yet it is the very essence of species richness.

The term *biodiversity* refers to this great variety of organisms, with special consideration to the interactions between these species and the environment.[13] The life and death of species, their origins and extinction, reflect the dynamic evolutionary spiral of biodiversity. If biodiversity represents the full matrix of ecological and evolutionary potential, then speciation and extinction are its agents.

One way speciation occurs is through separation in time and place. As climates, geographies, and habitats change, as biogeographic territories separate and converge, genetic configurations serve as specific adaptations to those fluid ecological and physiographic regimes. Some species differentiate so as to become genetically distinguishable. Over time, new branches grow on the evolutionary tree.

These same dynamic earth system changes, whether they occur on isolated mountain tops, in remote lakes, or over entire hemispheres, may change the environmental conditions so as to make life intolerable for any given species, leading to its demise and eventual extinction, severing the branch on the evolutionary tree. Extinction connotes the termination of evolutionary possibility for any species.

The ecological situation in any given place in time is a temporary record of this speciation/extinction process. Although sensitive observers can always detect resilient ecological and evolutionary patterns—configurations of life and landscape, or coherent assemblages such as biomes, ecosystems, and communities—what is most striking about biodiversity is the ever-changing fluidity of places and life histories. Every species has its own story and lineage. Every community has a different structure and history. Every landscape is unique. Biodiversity is the most tangible, ecological statement of inexhaustible biospheric creativity. And human beings have only glimpsed the smallest portion of this magnificent heritage. Extraordinary discoveries await anyone who is willing to explore the rich details of any living landscape.

Wilson conveys the richness of this heritage by likening biodiversity to books in a library. He suggests that one page containing vital natural history information be allotted per species. "If published in conventional book form, with pages bound into ordinary thousand-page volumes 17 centimeters wide inside cloth covers, this Great Encyclopedia of Life would occupy 60 meters of library shelf per million species." One hundred million species would require six kilometers of shelving, the equivalent of a medium-sized public library. [14] Surely this would be the mother of all field guides. Expand this encyclopedia so as to cover each species in more depth, or extend coverage to include other eras of earth history—paleoenvironments—and you will end up building an enormous city filled with thousands of libraries.

Paleontologist Richard Fortey, in his delightful book *Life: A Natural History of the First Four Billion Years of Life on Earth,* is intrigued by the "fateful combination" of tectonic plates, climate change, and physical geography. In discussing the archaic "supercontinent" Pangaea, he wonders what was beyond this land mass—an enormous ocean, filled with small islands—now entirely vanished, its traces covered by the movement of plates, bereft of fossil evidence or tracks of any kind. He imagines this lost hemisphere:

Who knows what strange Pangaean mutants might have signaled to their mates upon shores of impeccably white coral sand? We can plant the uplands with all manner of gorgeous and improbable plants of our own devising: ferns as big as palm trees, perhaps, for windblown fern spores would surely have reached these mysterious shores, We can imagine a perfumed breeze blowing across translucent lagoons, while some cumbrous crustacean—as improbable in its way as a Pacific land crab—picked a fastidious path on jointed legs through the litter of a silent forest. Maybe, even then, there were giant tortoises, or perhaps some strutting reptilian dodo, as grotesque as anything devised by Hieronymous Bosch. This forgotten oceanic world would have been populated by the vagaries of chance, blossomed through the opportunities that luck created, and had its evidence destroyed by another twist of circumstances. This lost hemisphere was a victim of the chanciness of things, played out on a globe of shifting plates. [15]

Fortey's lost hemisphere, constructed through scientific imagination, but entirely feasible, may represent missing volumes in the Great Encyclopedia of Life, or perhaps they are located in the fiction section under imagined ecologies. Our city of libraries resides as a metaphor in the infinite pathways of biospheric creation. Volumes are written, published, filed, retrieved, and borrowed, corresponding to the waves of speciation and extinction. The earth system sciences provide a tablature for organizing the libraries, speculating as to which volumes should be filed together. But the entire catalog is beyond human grasp, its plenitude akin to the multiplicity of biodiversity. Here are lost hemispheres, archaic habitats, extinct flora and fauna, all of which could be rearranged depending on dozens of biospheric variables. At any given moment in the history of the biosphere, the Great Encyclopedia of Life will have different volumes, and be subject to different arrangements,

Some scientists speculate that there is no upper limit to biodiversity. Physical geographer Richard Huggett suggests that various "climatic and geological processes incessantly increase the complexity of the physical environment—they drive geodiversity to ever-greater levels."[16] More habitats create more opportunities for specialization and speciation. Biodiversity is a dynamic, immanent, timelessly unfolding process. New life forms continuously emerge. The Great Encyclopedia is forever expanding.

Yet biodiversity is limited both by "background rate extinctions," or what may be described as the average, expected extinction rate given the ubiquity of local environmental change, and "megaextinctions," catastrophic species die-offs catalyzed by extraordinary global environmental change. These include a range of interrelated factors: cosmic causes (comets, radiation, solar flares), geological causes (continental drift, volcanism, sea level change, salinity change), and biological causes (spread of disease and predators, evolution of new plant types, change in biogeochemical cycles). The fossil record depicts five previous mass extinctions, three of which entailed the loss of more than 60 percent of all species. It is not known whether these extinctions were gradual or abrupt, and the chain of causation is subject to much scientific controversy and scrutiny.[17]

These five major extinction events represent an astonishing decrease in the diversity of life forms on the planet. In each case, a complete recovery, the reestablishment of previous species numbers, required tens of millions of years. As Edward O. Wilson warns, "these figures should give pause to anyone who believes that what *Homo sapiens* destroys, Nature will redeem. Maybe so, but not within any length of time that has meaning for contemporary humanity."[18]

Among conservation biologists, there is a virtual consensus that the earth is in the early stages of the sixth megaextinction. As a result of global economic development, dramatic population increases, urbanization, and natural resource use, not only are diverse habitats threatened but also the degradation of entire ecosystems is possible. The consequent fragmentation of the landscape is at the core of the current "biodiversity crisis."

Using the fossil record, scientists who calculate the mathematics of extinction estimate that in a biosphere of 5 million species, approximately one species goes extinct every two years. Assuming for estimations and variables, it's theorized that the annual average background extinction rate may range from one to five species. It's hard to estimate such rates because we are only working with fossil evidence, linked to probability and statistics. Nevertheless, there must be some baseline figure to which we can compare contemporary extinctions. Such tangible evidence relies on the cold, hard facts of species disappearances, yet estimates of annual rates can only be approximated, given that we still can't identify the number of species currently living in the biosphere, and that most extinctions pass unwitnessed.

Among the most widely cited figures, those which are prepared by estimating global habitat losses, there are some shocking, depressing numbers—between four and twenty-five thousand species per year are lost, which translates further to one to three species every hour! As you read this passage, another species has been lost. Andrew Dobson assesses these figures to mean "present extinction rates exceed speciation rates by a factor of around one million."[19] This mind-boggling figure far surpasses the estimated rates of even the previous five megaextinctions.

Such pronounced extinction at the hands of human expansion is not just specific to global industrialization. The so-called Pleistocene overkill, or the extirpation of the charismatic megafauna of North America and Europe during the retreat of the last glaciers, has been attributed both to climatic change and human diasporas. The human settlement of the Pacific Islands resulted in equally radical extinctions. Here are just a few figures to consider. Two thousand species of Pacific Island birds have gone extinct since human colonization began several thousand years ago. In the last one thousand years Madagascar has lost a dozen species of lemurs, several species of land tortoises and a species of hippopotamus. In the last ten thousand years, North America has lost nearly half of its endemic mammal species.[20]

Humanity's presence looms large. Whether at the hands of the hunt, the farm, or industry, whether in the Pacific Islands, the tundra, or coastal New England, the Pleistocene record is clear—the projected sixth megaextinction results from human activity. At various times in diverse locations, indigenous cultures have observed earth-based rituals, incorporating what we would now describe as ecological awareness. But more often than not, sometimes willfully but typically out of ignorance, human zeal overwhelms the living landscape, spelling trouble for the endemic ecosystem. With the unbridled power of economic development and the exponential increase of population, human impact (whether naive or malevolent) dramatically increases the rates of extinction, threatening biodiversity.

Biodiversity studies at once portray the complex magnificence of ecological creativity while proclaiming the prospects for its demise. This dual message permeates the psychospiritual atmosphere of conservation biology, field ecology, and environmental science at large. Interpreting the landscape so as to appreciate biodiversity requires a full-fledged perceptual commitment—using your senses to observe the abundance of life—as well as active scientific engagement, to understanding the ecological and evolutionary structure of the natural world. As a scientific learning process, biodiversity studies promote ecological knowledge. From a phenomenological perspective, you learn how to open your senses to the multiplicity of biospheric creation.

Whoever observes the natural world is also likely to notice threats to biodiversity. In contemplating biodiversity, you acknowledge the inevitability of extinction. An informed observer understands that extinction and speciation run their course with or without humanity, but when humanity serves as an agent of mass extinction, one has to deal with an awesome responsibility. Your actions, taken as a whole, in ways that you can't completely understand, have long-lasting impacts, on the order of tens of millions of years.

Who bears witness to extinction? Who feels the emptiness, the sadness, and the incomprehensible sense of loss when the last individual of a species expires? Typically there are no human witnesses. This enhances the mysterious, existential quality of the idea of extinction. You may viscerally observe the decline or demise of a species to which you have the privilege of relative proximity. But knowledge of extinction is mainly gained abstractly, by reading accounts of species losses around the world, or by grappling with the cold, stark mathematics of extinction statistics and extrapolations. The final act almost always occurs without human witness. Consider Edward O. Wilson's lyrical commentary:

Extinction is the most obscure and local of all biological processes. We don't see the last butterfly of its species snatched from the air by a bird or the last orchid of a certain kind killed by the collapse of its supporting tree in some distant mountain forest. We hear that a certain animal or plant is on the edge, perhaps already gone. We return to the last known locality to search, and when no individuals are encountered there year after year we pronounce the species extinct. But hope lingers on. Someone flying a light plane over Louisiana swamps thinks he sees a few ivory-billed woodpeckers start up and glide back down into the foliage . . . but it is probably all fantasy.[21]

This profound and forlorn passage captures the paradox of extinction. Only those who develop intimate knowledge of a species can bear witness to its loss.[22] With intimacy one cultivates an appreciation for the life history of a species, its unique ecology and behavior, providing the details for its description in the Great Encyclopedia of Life. What you don't know about you may never miss. It takes great courage to search for the last of its kind.

Most species pass away unnoticed. Many go extinct before they're ever named or identified by human taxonomists. Does that make their loss any less significant? Why should any one species matter more than any other? Why should a species be mourned—because of sentimentality, because it represents the end of evolutionary possibility, or perhaps because it is a statement about the human condition?

In their urgency to convey the importance of species extinction, conservation biologists will appeal to sheer self-interest, trying to explain that human survival depends on the integrity of the global environment. The prospect of mass species extinction spells doom for humans too. Niles Eldredge asks: "Why not just let the Sixth Extinction runs its course? After all, evolution ultimately creates new species that become the players in newly rebuilt ecosystems. The answer is simple: New species evolve, and ecosystems are reassembled, only after the cause of disruption and extinction is removed or stabilized. In other words, *Homo sapiens* will have to cease acting as the cause of the Sixth Extinction— whether through our own demise, or, preferably, through determined action, before evolutionary/ecological recovery can begin. Our fate is inextricably linked to the fate of Earth's species and ecosystems."[23]

So the fate of species is both the measure of biodiversity and a barometer of human destiny. Many fine ecologists, ethicists, theologians, and all manner of environmental thinkers have written superb rationales for promoting biodiversity—with reasons ranging from the economic and biological to the aesthetic and spiritual.[24]

What I want to emphasize here is the importance of bearing witness to extinction. One most appreciates life in the shadow of death. To observe the destruction of any ecological habitat or to perceive a threat to its integrity is a reminder of the fragility of biodiversity—how thousands of years of ecological and evolutionary history can be dispatched with a bulldozer, before you have a chance to even figure out what's happening. Habitats and species disappear overnight.

Perhaps there is a metaphorical correspondence between vanished species and "disappeared people." In a remarkable novel about the fascist regime in Argentina, the one famous for kidnapping dissidents in the middle of the night, Lawrence Thornton follows the visions of his protagonist Carlos Rueda, who is obsessed with the fate of the vanished. For reasons unknown to him, Carlos receives images and visions of what happened to these people—if they're alive, where they're located, and how they are imprisoned. He meets with the wives and mothers of the vanished to provide them with news of their loved ones. He realizes that his ability to imagine the whereabouts of the vanished kindles the flame of hope. As long as the fascist regime is unable to extinguish imagination, it cannot take over the hearts and minds of its victims.[25]

Species extinctions resemble the fate of disappeared people. With imagination you can recreate the circumstances of their absence—faces either familiar or unknown, in habitats both intimate and unseen. The hollowness of extinction serves to illuminate the fullness of creation. In the shadow of death, one must imagine life. In the next section, we consider how the experience of wonder provides the foundation for both hope and imagination.

Wonder and Indifference

In chapter 1 I described a trip to the winter migratory site of the monarch butterfly, high in the volcanic mountains surrounding Mexico City, in the groves of Oyamel fir trees. The plight of the monarch not only demonstrated the complexity of understanding global environmental change but it provided a way to reflect on the existential implications of the biodiversity crisis. In observing the butterflies, you are transfixed by wonder, but you also must contemplate the prospect of their demise.

While observing a meandering monarch in search of milkweed in your backyard, you might not know that over several generations it travels from the recesses of North America, from as far away as Canada, and eventually makes its way to the heart of Mexico. This genetic migratory behavior is surely one of the most extraordinary insect journeys. On frail wings and dynamic winds a monarch traverses unthinkable distances.[26]

Visiting the monarch reserve gave me an opportunity to contemplate this remarkable journey and to do so with butterfly experts, naturalists, and writers. It is a long trip from New Hampshire to Angangueo, Mexico. After several turbulent plane rides, long delays at airports, Mexico City traffic, impatient Mexican drivers for whom tailgating and dangerous passing is a form of folk sport, and harrowing, unbearably dusty mountain roads, I arrived at the sanctuary enthralled, but hopelessly dizzy. Yet my journey, for all of its psychological thrills and traumas, was less hazardous than that of the butterflies who survive dangerous perils and display far more grace. Every monarch butterfly east of the Rockies, and possibly some from the west too, makes this trip.

There are fifteen known winter monarch residences in this region, all of which are surrounded by encroaching agricultural, housing, and logging developments. Some are designated as "reserves," but even the so-called sanctuaries are threatened by unauthorized logging. On any given weekend day in January or February, thousands of tourists make a pilgrimage to see the butterflies. Busloads of people come from Mexico City on roads that can't really accommodate them. The sanctuaries are overrun.

I visited one of the better-organized sanctuaries, El Rosario. The trail entrance is lined with over a hundred shops selling food, trinkets, and all sorts of monarch souvenirs. This is supposedly a form of ecotourism, a means to support the local economy. Ecologically minded visitors are urged to unload their pesos here so as to endorse trinket sales over logging. An interesting tradeoff, for sure. A guide accompanies you on the steep path though the forest. You never feel as if you're walking through a wilderness. Rather, the impression is like being in a popular park in the coast range above Los Angeles.

After about an hour of hiking, you enter the realm of the monarchs. At first, you don't notice anything different, except a thicker, darker forest. Then someone points to a large oblong mass, drooping off a tree. You notice several of these masses, and then dozens. On closer glance you realize that these structures are the forms of hundreds of huddled monarch butterflies, nestled together for evening warmth in the high mountain forest. You look through binoculars and detect the details of this beautiful creature, forming a dense colony with its kind. You realize that the air is filled with butterflies, dropping from the sky after a feeding foray, preparing for sunset. Orange wings flutter throughout the forest. You observe the enormity of the colony. There are thousands and thousands of butterflies in this special grove of Oyamel firs—butterflies from all over North America. The butterflies drip from the trees into your heart. Bill McKibben suggests it is like "looking into the Mind of God."[27]

You are elevated above the turmoil leading to and surrounding the sanctuary. In observing the orange brilliance of the monarchs, in contemplating their remarkable passage, in the presence of their awesome confluence, in considering their fragility and resilience, you are overwhelmed with wonder. Abraham Joshua Heschel, the great Jewish theologian, described wonder as "the aboriginal abyss of radical amazement."[28]

In those speechless moments, when you're surrounded by such grandeur and fragility, you feel as if you are bearing witness to the magnificence of creation. You gaze through the aboriginal abyss. The living landscape swallows you whole. It becomes an enormous gap, deeper and wider than space and time. It's the place from which you originate. There is no need for explanation. There is only this ineffable experience. It's glimpses such as these, inexpressible as they are, that provide the deepest context for environmental learning. Concepts like ecology, evolution, and biodiversity merely hint at this radical amazement, giving you a way to express yourself. The thousands of visitors to the monarch sanctuary, most of whom lack any training in the environmental sciences, are pilgrims in search of inspiration.

What happens to these visitors when they return home to Mexico City or some far-flung place around the globe? Are their lives forever changed? Their visits (like mine) may be part sightseeing, part pilgrimage. In ways fully intentional, or undeveloped and nascent, they may respond to their monarch experience with wonder. Unless their visit represents the crassest form of tourism (I can't imagine that it does, as there are far easier and more comfortable places to sightsee), I find great hope in the event of these pilgrimages. They represent nothing less than a search for wonder. By visiting the reserve people make a statement about what's important to them. They are amazed and challenged. They're searching for meaning, looking to the butterflies for inspiration.

The relationship between biodiversity and wonder is the most pressing concern for any educator or citizen who seeks to address the sixth megaextinction. A state of wonder is the basis for an ethic of care. For decades, all manner of environmental educators, conservation biologists, and field ecologists have been striving to find ways to cultivate wonder in regard to natural history. Why doesn't everyone have the same sense of reverence and commitment to observing biospheric creation in this way? The most inspired environmental literature is the odes and reveries to living landscape, the testimonies to grandeur. How else do you wake the people up? Today, for educators who want to raise awareness regarding global environmental change, the challenge is how to cultivate wonder in the presence of the biosphere.

This challenge of educating for wonder is not merely the province of environmental studies. It is the prevailing challenge for any teacher who wishes to convey appreciation and reverence. Heschel bemoans its relative absence:

The awareness of grandeur and the sublime is all but gone from the modern mind. Our systems of education stress the importance of enabling the student to exploit the power aspect of reality. To some degree, they try to develop his ability to appreciate beauty. But there is no education for the sublime. We teach the children how to measure, how to weigh. We fail to teach them how to revere, how to sense wonder and awe.[29]

Can one even teach wonder? Or is it a quality that you exude? And if it is a quality, as unteachable and mysterious as the feeling itself, then perhaps it can only be modeled and reflected. The issue isn't so much to teach wonder as it is to exude it, to let it emerge unencumbered, to value its expression, to set up experiences where its manifestation is clear.

The role of wonder as an inspiration for learning has been the source of much ambivalence for scientists. In *Wonder and the Orders of Nature,* Lorraine Daston and Katherine Park, historians of science, provide an interesting history of the idea of wonder in relationship to learning about the natural world. Wonder at various times has been perceived as "a prelude to divine contemplation, a shaming admission of ignorance, a cowardly flight into the fear of the unknown, or a plunge into energetic investigation."[30] But since the Enlightenment, "wonder has become a disreputable passion in workaday science, redolent of the popular, the amateurish, and the childish."[31] Expressions of wonder among scientists are now reserved for their personal memoirs. Heschel warns that "the sense of wonder must not become a cushion for the lazy intellect."[32] Perhaps wonder and doubt together are an appropriate balance, a means of being simultaneously engaged and detached. Again, Heschel: "He who is sluggish will berate doubt; he who is blind will berate wonder."[33]

Heschel is writing in the context of theology and he is interested in the legacy of wonder that religious tradition contains. For our purposes, the issue is the legacy of wonder contained within the biosphere. For Heschel, the theological challenge is how to sustain the quality of wonder. Environmental educators face the same challenge—how do you make the daily experience of observing nature the source of wonder? Heschel suggests "since there is a need for daily wonder, there is a need for daily worship."[34] In Judaism, as in many traditions, a sense of wonder is maintained by uttering prayers and blessings of thankfulness, such as the prayer before the consumption and enjoyment of food. "Each time we are about to drink a glass of water, we remind ourselves of the eternal mystery of creation."[35] Here, too, is a means to appreciate the biosphere. With each drink of water, you can ponder the hydrological cycle as a biospheric process (see chapter 5). With every sighting of a monarch butterfly you can contemplate its remarkable migration story.

Yet the routine of daily life often restricts these sentiments. How do you worship what you've come to take for granted? "Life is routine," Heschel reminds us, "and routine is resistance to wonder."[36] For Heschel, praise and reverence sacralize the routine. This is accomplished through prayer and deed.

I wish to suggest that the same cycle of awareness is true for perceiving global environmental change. It is through the routine day-to-day observation of the natural world—the weather systems, the gardens, the butterflies, the pollinating insects, the wetlands—the list is endless—that you learn to bring the biosphere home. This is not accomplished through casual observation. It requires the perseverance of science and the ritual of practice—daily habits ritualized so they can serve as the source of both patient observation and unsurpassed wonder.

Yet how often do these daily events transpire unnoticed? You fail to pay attention to the butterfly in your garden because the phone rings, or you have something else on your mind. Or one's life circumstances are such that the daily economic or psychological struggle for survival overwhelms the legacies of wonder. Whatever your circumstance, surely you can reflect on the countless examples of your indifference. Heschel's comments on this are provocative: "Indifference to sublime wonder of living is the root of sin."[37]

No matter how much you strive, it is hard to imagine sustaining a state of perpetual wonder. Yet we have all encountered (or perhaps have been there ourselves) states of chronic indifference. Apathy breeds callousness. A degraded habitat withers from indifference, not wonder. What is more discouraging than to contemplate a home uncared for, a landscape unnoticed, an ecosystem unobserved, or a biosphere unappreciated by those who most benefit from its use? What is more depressing than encountering a bored adolescent, or a person, who says, "I have nothing to do?" How can this happen in a world filled with perpetual wonder? The monarch butterfly won't perish from too much love. Rather it will disappear out of callousness. Indifference to creation is the root cause of human engendered-extinction.

My visit to the butterfly reserve brought me to the "aboriginal abyss of radical amazement," but the journey also crossed a peninsula of foreboding. These reserves, islands in a sea of encroachment, face numerous development threats. The forest rapidly recedes in the face of surreptitious, midnight logging, highly eroded mountainsides, and the various development pressures cited earlier. Lincoln Brower and Robert Pyle, monarch researchers and naturalists who have visited the reserves periodically over the years, are dismayed at the signs of impact. On successive visits, they've observed the extraordinary reduction of Oyamel fir habitat. Monarchs face threats in the North as well, with the genetic manipulation of corn and the eradication of milkweed. Like the case of migrating songbirds, good conservation practices must be upheld at all the residences and waystations of cosmopolitan species. Brower and Pyle speculate that at current projections, the monarch butterfly may face its doom within a few decades.[38]

To stand in the monarch reserve amid the splendor of these extraordinary creatures is to open your heart to wonder, summoning the deepest reverence and praise. The grandeur of the monarch butterfly is enhanced by the fragility of its journey and the precariousness of its plight. To consider its demise is to feel great sorrow. Its glory is enshrouded by a disturbing halo. In contemplating creation and extinction from a naturalist's perspective, you experience both wonder and sorrow. The monarch's resilience embodies hope, its fragility inspires care, and its ecological prospects promote foreboding.

Hope and Foreboding

On returning home from the monarch journey, I was thrust into the middle of a local environmental controversy. The conservation commission in the town where I live recently completed a survey of the most fragile wetlands in the area. After careful consideration and numerous consultations with local experts, they decided to recommend several prime wetlands designations. This means, in the state of New Hampshire, a higher level of protection than is typically afforded by town regulations. But like most issues of this sort, local landowners who live adjacent to the wetlands feared that such a policy might restrict their autonomy and

used the issue to proclaim the importance of local control, the ineffi-
ciency of state bureaucracy, and the sanctity of their rights and responsi-
bilities as landowners.

At a recent town meeting, this issue was brought to a public vote.
Should the conservation commission go the next step in recommending
this designation to the state of New Hampshire? The public discussion
was heated as many landowners voiced their disapproval. As an alter-
nate on the conservation commission, a community member and a local
environmental thinker, I was compelled to advocate for the prime
wetlands designation. I had always kept a fairly low profile in town,
hadn't attended many public meetings, and never spoke up at town
meetings.

But in this case, I chose to say my piece, and I did so by referring to
my recent visit to the monarch butterfly reserve, using it as a means to
show the global consequences of a local action. The gist of my argument was that in less than a generation a forested landscape in Mexico
was deforested and that although the circumstances in my town and
Mexico were much different, without long-term protection and regula-
tion of vital habitats, open space could easily disappear. Implicit in the
message was a warning to take heed. What I had to say was no different
from the various projections and warnings that environmentalists the
world over customarily spout. The Mexican experience seemed somewhat far-fetched to me, but I felt it was a compelling way to place my
town's local environmental issue in a global perspective. I tried to
bring a biospheric orientation to our local decision, suggesting that in
protecting local wetlands, we were contributing to biodiversity and
species protection. Alas, my plea drew applause and support, but it
was not enough to change the outcome. The measure failed by just a
few votes.[39]

In effect, during that incident, in front of friends, neighbors, and ac-
quaintances, I played the role of biospheric advocate, asking people to
consider a future prospect that was not very appealing, the deterioration
of the local aquifer. I did so by describing the sense of wonder inspired
by the monarch butterflies, the miracle of biodiversity, and by reiterat-
ing the depth of appreciation I know that most of the town's residents
have for the place where they live. Yet I also asked folks to travel a darker
side with me for a while, to consider what could happen to local and
global biodiversity in the absence of strong regulatory protection. I was
not comfortable doing this, serving as a harbinger of doom, and any

eloquence I might have summoned may have been transcended by my self-consciousness.

Whether you patiently explain to a friend the dangers of global climate change, or teach a class about biodiversity and the sixth megaextinction, or engage in local activism to protect endangered species and habitats, you are serving as a biospheric advocate. Using informed speculation, employing the "facts" of ecology and global change science, you take your friends, students, colleagues, neighbors, and political adversaries on a trip to territory they might prefer not to visit. From my experience, this is the single most complex challenge in teaching about global environmental change. To address an issue with full urgency requires a good measure of foreboding, but to sustain commitment and resolve, one must summon hope. Balancing these dimensions of the human experience is every bit as much of the global change curriculum as learning environmental science. Many prominent global change scientists, ecologists, naturalists, and conservation biologists at various times in their public lives take on the role of biospheric advocate.

In his two-volume study, *The Prophets,* Heschel is concerned with the role of the prophet as the conveyor of moral, political, and spiritual change. Biblical prophets gather notoriety via warnings of doom. Citing corruption, depravity, injustice, and lack of attention to the word of God, they become a conscience, and perhaps a nuisance. But they also provide a message of hope, holding out the prospect of redemption and resurrection, if the people reform.

Prophets "disclose the future in order to illumine what is involved in the present."[40] They elaborate on the trends and patterns of the contemporary era. The biblical prophet serves as God's voice, conveying His concern and judgment. Hence the prophet lives at the edge of time, using history as the means for proclaiming God's eternal message. The biblical prophet derives inspiration, vision, and instruction from God. "His true greatness is his ability to hold God and man in a single thought."[41]

It is presumptuous to compare environmental activists and educators to biblical prophets. Rather, I suggest that there are situations when people who speak publicly about the loss of biodiversity, species extinction or global warming take on a prophetic role in conjunction with their advocacy. As such it is helpful to understand the role of the prophet as a visionary messenger. Biospheric advocates, unlike biblical prophets, rely on scientific research and projections rather than the word of God.[42] Informed by evolution, ecology, and the earth system sciences, they convey an environmental message about the future. Biospheric advocates live in deep geological time. They interpret the contemporary Pleistocene epoch with the full evolutionary and ecological spectrum in mind. Their dire prognostications are based on their observations of ecological phenomena. With the fate of the earth at stake, they serve as messengers of the biosphere.

For the prophet, "every prediction of disaster is in itself an exhortation to repentance."[43] The biblical prophet is concerned about moral failure and spiritual decline. The biospheric advocate is worried about the abnegation of ecological responsibility. A moral meaning is implicit in this message. Heschel writes that "the prophets remind us of the moral state of a people: few are guilty but all are responsible."[44] This is exactly the message of many environmentalists. Ecological deterioration is everyone's responsibility. We must change the way we live. For the biblical prophet, every thought, intention, and action must be scrutinized as if it is taken in God's presence. For the biospheric advocate, one's daily habits and actions must be taken with the biosphere in mind.

Heschel writes that to be immersed in the prophet's words exposes you to the "ceaseless shattering of indifference" and "one needs a skull of stone to remain callous to such blows. I cannot remain indifferent to the question whether a decision I reach may prove fatal to my existence."[45] Every action can change the direction of history. For the biblical prophet, as long as there is injustice, suffering, greed, and corruption, there can be no rest. For the biospheric advocate, every assault on biodiversity is cause for concern. Any extinction can change the future of an ecosystem. The prophet and advocate both espouse constant vigilance. Every conscious moment is filled with a visionary directive.

For Heschel, sustained indifference unleashes history's darkest hours. His rabbinical and philosophical studies took place in the deep shadow of the spread of European fascism. He fled Germany in 1940, observed (at a distance) the destruction of six million Jews, and lost most of his family in the Holocaust. His major theological statements, *God in Search of Man* and *Man Is Not Alone,* were written in the early 1950s during the dawn of the atomic era. His work reflects a perennial tension between a phenomenologically oriented theology based on wonder, awe, and "radical amazement," and the shadow of unspeakable evil. Although he rarely addresses the Holocaust specifically, the question of good and evil permeates his work, informed by the direct experience of the most dreaded horrors of the twentieth century.[46]

As Heschel's theology contrasts reverence and praise with unspeak-able evil, it is of great interest to anyone who contemplates creation and extinction in the context of biodiversity. Comparing the annihilation of six million Jews to gross species extinctions may be of a different moral order. I leave that for the reader to decide. But I am convinced that the terms and conditions of Heschel's spiritual development are of seminal importance for keeping global change in mind. How does one cultivate wonder, summon praise, and exude reverence in the face of unthinkable suffering and sorrow?

Prophets preach vigilance because they see how much has gone wrong and they anticipate the consequences. Indifference may spread unnoticed until the whole house comes tumbling down. "It took one storm to turn a civilization into an inconceivable inferno . . . Only fools are afraid to fear and to listen to the constant collapse of task and time over their heads, with life buried beneath the ruins.[47] Understanding the full measure of indifference leaves you with feelings of dread and foreboding.

If the prophet's message seems extreme, it's because people grow im-patient when they are constantly reminded of their excesses and negli-gence. The prophet's standards of spiritual behavior may seem too demanding to uphold. The biospheric advocate's standards of environ-mental behavior may seem equally unreasonable. And people may grow wary of false prophets or self-righteous advocates who use spuri-ous and grandiose claims merely to gratify their own ego. Prophets and advocates are both judged harshly, as they should be.

What Heschel is getting at is that you must juxtapose the ideal of per-petual wonder with the reality of relentless vigilance. Just as every mo-ment is replete with wonder, so is every moral action pregnant with significance. There is no room for indifference. A balanced vigilance is required. Daily worship is the seed for cultivating perpetual wonder. Moral grandeur is the means to overcome indifference.

In learning to interpret global environmental change, one is bound to contemplate hope and foreboding. Biospheric perception inspires radi-cal amazement. But the possibilities of species extinctions and environ-mental degradation loom equally large. Indeed, these observations must be balanced accordingly; otherwise you live in a dream world or a state of chronic despair. Heschel's theology prescribes a middle ground. It is the very tension between these feelings that promotes both environmen-tal and spiritual learning.

For Heschel, "endless wonder is endless tension, a situation in which we are shocked at the inadequacy of our awe, at the weakness of our shock, as well as the state of being asked the ultimate question."[48] To dwell in wonder is to contemplate questions of meaning and purpose. "Wonder is the state of our being asked."[49]

After visiting the butterfly reserve, our group of writers, educators, and scientists pondered exactly this question. Our common sense of wonder in the presence of the monarchs prompted a lively conversation. Wonder inspired curiosity—how did these butterflies come to be here? It broadened scale and perspective—the world is so much bigger than you can ever imagine. It placed the human condition in an ecological and evolutionary perspective—the migratory accomplishment of the butterflies is remarkable. We appreciated the sheer aesthetic quality of the experience. It was very beautiful to see. And we thought about the future—what will happen to these butterflies? But one question prevailed above all others—what does this experience demand of us? What is it asking us to do?

Wonder Spawns Indebtedness

Whether it is in the presence of the monarch butterflies, the spring peepers, this morning's birdsong chorus, or the howling wind during a driving rain—in cultivating wonder you are filled with praise. What is praise? It is feelings of appreciation, gratitude, humility, and reverence, the sense that you are witnessing a complexity and richness of life experience that you can barely grasp.

Praise isn't premeditated. It's not something you think about in advance. It just comes out of you like a song or a poem, as if you are merely a temporary voice for creative expression. Whatever it is you say, sing, write about, or create through such expression is a means for the biosphere to exalt itself. Religious traditions summon praise through blessings and prayer. Scientific traditions summon praise by studying a phenomenon with the most rigorous measure of dedication and intellect. Artistic traditions call forth the creative imagination. The shape of praise takes many different forms and its object reflects infinite conceptions.

Why is praise important? It engenders appreciation and gratitude, a thankfulness regarding what you've been provided. Praise the rain, the soil, and the sun for this morning's breakfast. Praise ecology and evolution for biodiversity and the migration of monarch butterflies. Praise the biosphere for planetary life.

Praise is the first phase of a response to wonder. What follows praise? Praise alone leaves you feeling empty, as if you've been given an extraordinary gift without responding in kind. But what exactly can you give back? What actions can you take to demonstrate your gratitude? Who is the recipient of such actions and what is the source of the demand? Wonder spawns indebtedness.

As we pondered the fate of the monarch butterflies, we realized that we were deeply indebted. The quality of our lives, the prospects for our future, our moral character, was reflected in the precarious condition of the reserve. What would human life be like without monarch butterflies, without migrating songbirds, without wild places? What political and economic situations compel people (in Mexico or the United States) to log Oyamel forests, to eradicate milkweed, to squander biodiversity?

What apprenticeship is worthy of our indebtedness? We are all familiar with the long list of noble possibilities—the arenas for political action, environmental education, ecological restoration, sustainable development—there are many virtuous paths to follow. But perhaps the greatest challenge of all is to find the perennial reminder of indebtedness in our daily actions, the willingness to owe rather than own, to give as well as receive.

While in Mexico our group of butterfly pilgrims stayed at a delightful inn on several acres of lovingly manicured landscape, an oasis amid the unbounded chaos of a country in rapid ecological transition. The proprietor and several locals told me that thirty years ago the surrounding countryside was a paradise—deep mountain forests and small farms. The inn bordered a small ravine and the adjoining hillside was filled with garbage, obviously emptied from the shanties on the other side. Like many places in the world, Mexico is a land of shocking extremes. There are places of unsurpassed beauty and uncommon community surrounded by unspeakable squalor. Their contrast is further reminder of the need to balance hope and foreboding.

As we were further immersed in our conversations about the plight of the monarch butterflies and all of the threats to its habitat, with shanties in clear view from the peacefulness of our bucolic retreat, a monarch fluttered down from the sky and landed on a bush in full display. We gazed at it interminably, enraptured in silence, transcending this moment to travel to all the places where this butterfly wanders.

This morning, the first warm day of early spring in southwest New Hampshire, as I am writing this essay, sitting on the porch, a mourning cloak butterfly, my first butterfly sighting of the season, lands in front of me. Not a monarch, but a butterfly nonetheless. I look at all butterflies differently now, each one a reminder of the broad ecological dimensions of the biosphere. But this butterfly summons much more than a heightened ecological awareness. It represents hope, the inimitable resilience of the biosphere, and the fluttering of praise. What does radical amazement and a sense of wonder ask of me? It demands that I cultivate hope. This is the venue for my indebtedness. In cultivating biospheric perception, you engage in an apprenticeship with hope.

An Apprenticeship with Hope

In his poignant and lovely book, *Hunting for Hope,* Scott Russell Sanders tells the story of a whitewater rafting trip he takes with his adolescent son. As they drive toward their destination, it is clear to Sanders that there's a great deal of tension between the two of them. He understands that some of the tension can be explained by their changing relationship. Fathers go through this when their sons become teenagers. But he thinks that there is something deeper and he probes to discover it. After some gentle, but relentless questioning, Sanders uncovers the issue. His son suggests that Sanders carries a shadow of gloom wherever he goes, his dismay about the future of the planet prevents him from enjoying anything. Moreover, Sanders' persistent, negative commentary on malls, advertisements, developments, and a host of assorted ills, are all judgments on the world that his son has to grow up in.

Sanders is deeply moved by what his son tells him and wonders if this is the message that he is projecting both to his children and his students. His son claims: "You make me feel the planet's dying and people are to blame and nothing can be done about it. There's no room for hope. Maybe you can get by without hope, but I can't. I've got a lot of living still to do. I have to believe there's a way we can get out of this mess. Otherwise what's the point? Why study, why work—why do anything at all if it's going to hell?"[50] Sanders realizes that although this was perhaps unfair, a caricature even, "there was too much truth and too much hurt in what he said for me to fire back an answer."[51]

Hunting for Hope is a beautiful meditation on where hope resides—in Creation, in wildness, in our bodies, with family, in fidelity, simplicity, beauty, and craft. But in cultivating hope, Scott Sanders has to come to grips with his dark visions, too. He realizes the discomfort that his prophecy causes. "I cannot scour from my vision the darkness that troubles my son, because I have witnessed too much suffering and waste, I know too much about what humans are doing to one another and the planet."[52] In this regard, he will forever rest uneasy. What's particularly interesting about Sanders's approach, and his effective use of memoirs, is his emphasis on memory and projection. "Hope and memory are kindred powers, binding together the scraps of time."[53] He summons hope for the future by resurrecting the most cherished moments of his past. This lays the groundwork for his most important insight, that most forms of therapy aspire to make peace with the past, but just as crucial is the necessity of making peace with the future. "If I am able to live fully in the here and now, I must be able to look both backward and forward with clear vision."[54] Hope's garden of meaning connects the past and the future.

In teaching about global environmental change, you learn that the most interesting and vital conceptual challenge is exactly this issue, how to connect time and space through one's observations of the past, the living experience of the present, and concerns about the future. The more you study the history of the biosphere and the evolution of life on earth, the more you realize the infinitesimal context of a single human life span. Contemplating 4.5 billion years of earth history, if nothing else, should provide a sobering dose of humility.

How then do you reconcile the inexorable impact of human expansion on the biosphere—humans as agents of the sixth megaextinction—given the seeming insignificance of any human action? In the very long run, what difference can any individual action possibly make? Here is a perplexing existential dilemma. In contemplating the biosphere you are impressed with the utter insignificance of your life, your total inability to grasp the full evolutionary measure of your existence, your complete and total naiveté when it comes to questions of meaning and purpose. Why bother? Why pay attention? This is not just a neurotic's rationale for apathy. It is a legitimate, understandable response to thinking about the scale of global environmental change. Ironically, this avenue, while it may inhibit activism, also reveals the arrogance of human agency. Even the most egregious development projects are short-term!

Yet it is the very incomprehensibility of the scale of your life and actions that is both a foundation of meaning and a source of hope. Just as you can't fully grasp the conditions of your existence, so you can't assess the long-term impact of your actions. Whatever you do, think, or say may precipitate a chain of events that is way beyond your ability to interpret or follow. The idea of karma is that every thought moment reverberates endlessly throughout time. You are always responsible for your actions, which are far more important than you can ever imagine. Here's the paradox—what makes you think that you're so important that you can understand the deep implications and ramifications of what you do? You can't take yourself too seriously, but don't ever underestimate the impact you might have.

The sense of being overwhelmed by the exigencies of events beyond your control emerges with most every discussion of global environmental change. How can the "ecologically minded" possibly temper the engines of global development? How can we stop the transformation of the landscape, the tide of history? These feelings of futility and uselessness, as understandable as they may be, and acknowledging the extent to which any activist or educator has them, are the call of the victim, a self-imposed powerlessness, and perhaps an abnegation of responsibility. How can you be so sure that you can project the flow of history and the course of events? How do you know what is going to happen?

Irving Greenberg, in his book *The Jewish Way,* comments on how this question of action is of special significance during the Jewish holiday of Yom Kippur. On this day, you consider through deep introspection your moral and spiritual behavior during the course of the year. It's a trial of sorts, in which all of your actions are judged in the context of eternity. Greenberg cites Moses Maimonides who comments on the cosmic significance of every action: "Everyone should regard himself throughout the year as exactly balanced between acquittal and guilt. So, too, he should consider the entire world as balanced between acquittal and guilt. If he commits one additional sin, he tilts down the scale of guilt against himself and the entire world and causes its destruction. If he performs one good deed, he swings himself and the whole world into the scale of merit and causes salvation and deliverance to himself and his fellow men."[55] The world is balanced on a fulcrum, and by virtue of your choices, it can swing in either direction.

Given the enormity of one's life choices and prospects and the limited awareness you have of your situation, it is unavoidable that one should experience both hope and despair. Hope emerges from being needed. Despair emerges from feeling useless. There is no escape from the tension between these impressions. There are infinite sources of hope and millions of reasons to plunge into despair. So it should be no surprise if your life sometimes feels like a tug and pull between these powerful forces. Perhaps the most balanced approach is to walk a fine line—to be inspired by hope and challenged by despair—and to do so with compassion and detachment.

In learning about global environmental change, you are exposed to the calamities of environmental degradation—pollution, species extinctions, habitat loss, and incomprehensible sorrow and suffering. Whether at the hands of globalization, war, or any assortment of human caused impacts, or by the vagaries of earthquakes, hurricanes, and the great panoply of earth system changes, this is the undeniable reality of the global situation. The matter is too urgent for despair. Rather, it requires vigilance, attention, and compassion. The detritus of indifference accumulates unnoticed until the sediment hardens, forming layers of callousness. It takes courage and commitment to combat callousness, both in the community and in your heart.

While watching butterflies, listening to peepers, or gazing at a cloudy sky, you also learn about the remarkable complexity and beauty of the biosphere. There's so much to study and learn about—the biogeochemical cycles, weather systems and patterns, biodiversity, the history of life on earth, the depth and richness of this extraordinary planet. What a privilege to be granted the gift of this learning. By studying the biosphere, you practice perception, bear witness to creation, and hold open the prospect of perpetual wonder. You learn how to experiment with radical amazement. This education calls for gratitude, praise, reverence and exuberance. This too requires an open heart and an attuned mind.

The foundation of learning about global environmental change is an apprenticeship with hope. The remainder of this book sets out some of the terms of this apprenticeship. The next two chapters integrate radical amazement with natural history observation. The first step in bringing the biosphere home is to be aware of its presence. You bear witness to the biosphere by developing intimacy and familiarity with the living landscape. Is it too simple to say that there is no solution to the biodiversity crisis unless we all pay attention? In *Hunting for Hope,* Scott Sanders reminds us that "Creation puts on a non-stop show," and that the challenge for anyone who seeks genuine hope is "to discover, or rediscover, ways of entertaining and sustaining and inspiring ourselves without using up the world."[56] Learning how to observe the biosphere is a good place to begin.

Another means to cultivate hope is through deeds and actions. You can't demonstrate gratitude and reverence through observation alone. It is through the substance of your relationship with family, community, and ecosystem that you encounter appreciation and humility. But where does one find intimacy and familiarity in a fast-paced world, where people and species are perennially on the move? What kinds of deeds endure? Chapter 7, "Place-Based Transience" suggests that whether you live in a place for one day or a hundred years, you still have a responsibility to give something of yourself in return.

Nourishing and sustaining hope is the most daunting challenge for anyone who is concerned about global warming, species extinctions, or any of the human and ecological consequences of these processes. Without doing so, it is impossible to keep global change in mind. The consequences are just too distressing. Hope is inspired by spring peepers, monarch butterflies, dazzling arrays of stars in the heavens, the miracle of biodiversity, and the warm glow of communities working together to appreciate and preserve their biospheric legacy.

4   A Place-Based Perceptual Ecology

Learning to perceive global environmental change requires a daily practice of natural history observation. You learn how to move from the immediate experience of familiar species and landscapes to broader biospheric perspectives of space and time. You start with the place where you live. You observe life forms, the landscape, the rhythms of seasons and sunsets, the daily weather patterns, and the soil—the rich tapestries of your local ecosystem. You pay close attention to the details and patterns that are right in front of you, what you experience on a day-to-day basis. Gradually you become familiar and intimate with the landscapes of home and in so doing you develop a foundation of knowledge grounded in tangible experience. This is the basis of local expertise.

The more familiar you become with the place where you live, the more you'll come to recognize the importance of the relationship between other places and your own. Inevitably questions will emerge as you try to understand why and how the local environment changes. These questions lead you to explore other places and to see how what happens locally often depends on a global context. Today's local natural history is an evolutionary response to yesterday's environmental conditions. You find that it is much easier to step out of place and time when you have your feet firmly placed in the here and now.

The paleontologist decodes ancient earth history through the message in a fossil. The geomorphologist reads the movement of ice in the placement of huge boulders. The community ecologist describes the history of the forest by reading the bark of trees. The microbial ecologist evaluates water quality by noticing which algae are in the lake. These are just a few of the many signs and texts that can be interpreted through skilled observation. This kind of interpretation reflects finely honed perceptual skills as well as the rewards of scientific training.

Such interpretive expertise is crucial if communities intend to deal with the complexities of global environmental change. This expertise is too important to be left in the hands of specialists. It must become accessible to all concerned citizens. We will only come to grips with the challenges of species extinction, habitat degradation, and global warming when entire communities mobilize to learn about these issues. Such learning is most meaningful when it emerges from tangible experience.

In this chapter, I develop the foundation of a place-based perceptual ecology, an approach to thinking about global environmental change that leads from the intimacy of local natural history to the biospheric dimensions of place and time. I begin by explaining what I mean by perceptual ecology and why I think it is such an important approach to environmental learning. Then I describe in more detail why a place-based orientation is an appropriate foundation for thinking about global issues. This opens the path to the heart of the chapter—exploring what the practice of perceptual ecology actually entails. Using local natural history as a guide and the life of the imagination as inspiration, I propose an experimental curriculum of scale, designed to detect local and global patterns within the perceptual limitation and potential of the human organism. Crucial to this approach is the juxtaposition of space, place, and time—being able to move adeptly among and between time frames, understanding the relationships between different places, and considering how places change through time.

A Perceptual Ecology

An etymological exploration of "perceive" reveals its most basic meaning—to apprehend through the senses. What can be more obvious than this? The most "common" sense is the shared assumption that people learn about the world through their five senses. Yet the exact process through which this happens, the way the mind sifts through its sensory impressions, the filters and windows that mediate those impressions—these are deep and complex issues, the source of many controversies in philosophy and psychology.

Most people learn about the natural world through what they directly perceive.[1] A beautiful sunset gets your attention. A cold, icy rain will make you very wet and uncomfortable. A field of blooming wildflowers is lovely to see and smell. All of these scenes evoke vivid sensory impressions, triggering memories and associations that reverberate through every corner of consciousness. You can't have any of these experiences unless you are outside to appreciate them. Buried deep in the recesses of a windowless office, you may never hear the thunderstorm. Staring at your computer screen, the subtle nuances of the daily weather may entirely escape your attention.

Yet these sensory impressions of nature are vital to human awareness. I am deeply moved by David Abram's call for a reciprocal relationship between human senses and the sensuous earth. "Humans are tuned for relationship. The eyes, the skin, the tongue, ears, and nostrils—all are gates where our body receives the nourishment of otherness. This landscape of shadowed voices, these feathered bodies and antlers and tumbling streams—these breathing shapes are our family, the beings with whom we are engaged, with whom we struggle and suffer and celebrate. For the largest part of our species' existence, humans have negotiated relationships with every aspect of the sensuous surroundings, exchanging possibilities with every flapping form, with each textured surface and shivering entity that we happened to focus upon."[2]

This type of intimate relationship demands proximity and attentiveness. You learn about the place where you live if you walk the fields and pathways, whether they are mountain trails or city streets. You learn the local birds if you get out and watch them, listen carefully to their songs, observe their behaviors. And you can do this whether you are perched at the edge of a remote forest or walking through Central Park. You rely on your senses to learn about the natural world, and in so doing, you develop intimacy with your surroundings.

Perceptual ecology suggests using the senses to apprehend the interaction of organisms with their environment. The study of ecology gives you a way to think about what your senses apprehend. Perceptual ecology has a reciprocal quality. The visceral impressions of the forest are virtuous both for the experience and its interpretation. It is important to ask good questions about what you observe. That is the heart of scientific inquiry. Why do these trees live here? What is the soil matrix? What is the role of the mychorrhizal fungi? How will global warming influence this habitat?

There are many ways to "interpret a forest." A particularly powerful approach is to merge sensory awareness with scientific inquiry so that both paths inform each other. To learn about a tree requires participatory learning. You touch, taste, and smell its bark. You trace the outlines of its twigs and leaves on your drawing pad. You feel its staunch majesty. This visceral experience requires an ecological perspective. Can you see the forest for the trees? To understand the tree in relationship to forest ecology entails more than sensory impressions (as important as they are). More can be learned about the natural history of the tree if you look it up in your field guide. With additional reflection you can ponder the tree, reading about its life history, its associated flora and fauna, its biogeography. Then you might consider the complexity of its ecology. What function does the tree play in the ecosystem? How does it cycle nutrients? Later we'll see why questions such as these are crucial for understanding global environmental change.

This melding of the senses and science is the crux of perceptual ecology. What you directly observe is amplified by a solid foundation in natural history, and given further grounding when supported by the framework of scientific ecology. The great naturalists share this fine sensibility—the ability to mesh sensory, aesthetic, and scientific observations for the purpose of environmental learning. This resounds through the work of Thoreau, Muir, and Carson, as well as many others. Hence I propose perceptual ecology as a means to cultivate sensory awareness, interpretive natural history, and scientific ecology. This is the basis of a reciprocal, participatory learning pathway. But we still have to add the next phrase to the expression. Why a "place-based" perceptual ecology?

Why Place-Based?

Sense of place is at the core of many environmental learning initiatives. It comes from an eminently practical premise. People are typically interested in understanding who they are in relationship to where they live. By exploring the places that are most important to them, they are more likely to take an interest in the human and ecological communities of those places. Exploring sense of place involves thinking about home and community, ecology and history, landscape and ecosystem. In essence, it is a search for your ecological roots, and a way to link your ecological identity to lifecycle development. I have written extensively about the importance of this concept for environmental education in my book *Ecological Identity*.[3] The tacit knowledge of environmental educators presumes that achieving a reflective sense of place contributes to an ethic of caring about habitats and communities.

Sense of place has appeal because it conveys feelings of rootedness and stability in a world of dynamic environmental change. It's a response to the demographic turbulence of people constantly on the move, the overwhelming complexity of issues on a global scale, and what seems like powerlessness in the face of cultural and environmental trends beyond the purview of any locality. Achieving a sense of place allows you to identify with the place where you live, to take responsibility for its quality of life, to become familiar and intimate with your local surroundings. The bioregional movement uses this emphasis on place and community as a means to construct approaches to environmental policy and issues of political economy.[4] A place-based orientation is appealing to educators and activists alike—it lends a certain tangibility to otherwise complex environmental issues.

My claim is that a place-based orientation is also the foundation from which to explore global environmental issues. First, I wish to offer a *cognitive rationale*. David Sobel, in his groundbreaking book *Mapmaking with Children: Sense of Place Education for the Elementary Years,* elaborates a developmental scheme for teaching geography to children. He describes how children develop a relationship with the natural world and social community from ages five through twelve. His comments on cognitive development are of great interest and relevant for all phases of environmental learning.

The progression of children's mapmaking skills is a microcosm of cognitive development in elementary school. At five and six, children are still immersed in early childhood and their world is small, contained, and dominated by sensory perceptions. The right hemispheric mode of spatial and visual perception dominates, and feelings and pictures are main forces in the organization of the child's world. The houses, trees, and animals are faces in the landscape that carry a certain emotional valence and their "look" needs to be preserved along with the relationship between the child and the aspects of his or her surroundings. By eleven or twelve, the child has gained perspective, both literally and figuratively. The child has gained the ability to illustrate a view from a chosen perspective—to stand back, separate from the scene, and observe it. While the younger child is bound up in the lack of differentiation between subject and object, the older child can take an objective look at the subject of the landscape. The evolution of children's maps can tell us how to approach geography and environmental education curriculum, and it can give us insight into the cognitive changes that characterize the challenge of moving from kindergarten to sixth grade.[5]

The practical core of Sobel's argument is that younger children who are still deeply embedded in a matrix of home, earth, and family should make big maps of small places. As they reach middle childhood and are more cognitively equipped to expand their territories of exploration, they are interested in and able to map bigger landscapes. At age 12, near the end of elementary school, they develop the ability to draw and experience panoramic views. There is a developmental progression in the cognitive capacity to distinguish various approaches to scale. Indeed, this is an area that begs for more research—at what point are advanced concepts of scale, especially as they apply to global environmental issues, developmentally appropriate (see chapter 8)?

Although these guidelines are designed for elementary school children, I suggest that they are a useful guide for perceptual ecology as well and a microcosm for environmental learning generally. Here's why. Perceiving global environmental change requires the ability to scan broad horizons of place and time. For example, species extinction and biodiversity are hard to grasp without an understanding of evolutionary biology. Global warming can only be understood in relationship to paleoclimatology. All three concepts involve some familiarity with the history of life on earth—plate tectonics, ancient landscapes and atmospheres, megaextinctions, and cosmic impacts. These concepts are highly abstract, intangible, and require the skilled juxtaposition of scale. They demand an advanced level of cognitive sophistication. Just as you wouldn't expect an elementary school child to understand these issues, it is somewhat overbearing to expect a layperson to jump into the conceptual maelstrom of global environmental change. Many environmental practitioners, people who are familiar with the complexity of these issues, also struggle with such conceptualizations.

Yet the challenge of environmental learning, indeed, the very premise of this book, is that perceiving global environmental change is the prerequisite for taking responsible action. I suggest that intimate familiarity with local natural history provides the ground floor for this challenge. You build a foundation of knowledge to which you can always return. You start with the place and time that you know best before you venture out around the globe. The same developmental sequence that young children use for learning about place is valid throughout an entire lifetime. You start by making a big map of the place where you live. Fill in the details of who lives where. Notice the intricacies of habitat and community, of seasonal variation, of microclimatic change. When you have enough experience to observe specific changes, then some broader patterns may begin to emerge. That's when a panoramic view may reveal an expanded landscape. The more penetrating questions you ask, the more changes you'll observe, and the farther you'll have to travel in space and time to get some answers. But your travels are always taken knowing that you have the conceptual territory of a local place to return to. Later in the chapter I explore this process in more detail. For now, let me assert that perceptual ecology requires local embeddedness, at a scale that is appropriate to how much a person can learn, to the limits of what one is able to take in.

One of the tenets of a place-based approach is that environmental awareness entails familiarity with the basic natural history of your home place. Often when I am in educational settings, I suggest that the question, "what do I know about the place where I live?" should be crucial to environmental learning. Recently I was asked by a local high-school teacher to speak to his philosophy class about Thoreau. I thought that as an entree to this discussion, I would ask the students some basic ecological and geographic questions about their bioregion, guided by the now classic "Where You At?" quiz, published many years ago in the *CoEvolution Quarterly*.[6] The quiz is designed to get you to think about your daily observations of the natural world, asking questions about weather, landscape, flora and fauna, environmental history, and geology. These students were juniors and seniors in a rural high school. They had all taken biology. I was shocked by their inability to answer these basic questions. They revealed an enormous hole in their natural history knowledge.

Gary Paul Nabhan and Sara St. Antoine performed a landmark study in which they surveyed the natural history knowledge of a young generation of O'odham and Yaqui Indians and contrasted this with the knowledge of elders in the same community. They found that acculturation, accelerated by the advent of television, linguistic assimilation, the disappearance of storytelling, and the decline in direct outdoor experience, led to a precipitous loss of natural history knowledge and awareness.[7] Robert Michael Pyle describes this as the extinction of experience, "simply stated, the loss of neighborhood species endangers our experience of nature."[8]

Speak to environmental educators who are on the frontlines, working in a classroom, park, or nature center, and their experience will confirm this trend. People spend far more hours in front of video screens than they do walking the land. It is only the unusual classroom or school that takes natural history and ecology seriously as crucial to the curriculum, or that stresses direct outdoor experience as intrinsic to learning about nature. Compounding this problem is what Peter Kahn describes as "environmental generational amnesia," the idea that people "take the natural environment they encounter during childhood as the norm against which to measure environmental degradation later in their life."[9] You can assess the cumulative result.

Although many young people (and adults) are aware of problems such as tropical rainforest destruction, global warming, and species extinction, and may show a great deal of concern about environmental issues, they are less likely to be intimately involved with the plants, animals, and landscapes of their home bioregion. Consequently, the relationship between global and local environmental issues appears vague and insubstantial. A place-based approach presents a hands-on alternative. You transform ideas that are vague and insubstantial into issues that are tangible and intimate. You do this via proximity and direct experience, through participatory learning.

The second rationale for place-based environmental learning is the *naturalist imperative.* This reflects an unabashed ecological urgency. You will never know what is missing from your home landscape until you have the natural history expertise to observe it for yourself. You will never attain the deep emotional connection and involvement with the bioregion until you can affiliate with its human *and* ecological community. This requires the skills of the naturalist. It is a prerequisite of understanding ecology, environmental history, biodiversity studies, and earth systems science. You prevent species extinction, in part, by restoring the knowledge of natural history.

Nabhan and St. Antoine identify three key strategies for "staving off the extinction of experience: intimate involvement with plants and animals; direct exposure to a variety of wild animals carrying out their routine behaviors in natural habitats; and teaching by community elders (indigenous or otherwise) about their knowledge of the local biota."[10]

Let us add to this what John Elder describes as eight "stories in the land," an educational approach that is "firmly rooted in our home landscape."

1. the geological processes producing the main landforms of our watershed;

2. its characteristic climate, seasons, and weather;

3. the life cycle and habitats of notable local plants and animals, and the communities to which they belong;

4. the fluctuations of forest history in our region;

5. indigenous human cultures, and some of their own stories about the area;

6. immigration and settlement from other countries;

7. nearby farms and their products;

8. ways people in the watershed presently make their living.[11]

Now the discussion of "place-based perceptual ecology" is more complete. It should be clear how intimate natural history knowledge, sensory awareness, and ecological interpretation form an interconnected, reciprocal, participatory, learning milieu. There is one more element to this. There is a *naturalist sensibility* that embodies this learning path. Before I describe the dimensions of a perceptual ecology practice and how it may expand to encompass global concerns, I will highlight the qualities of this sensibility, and consider why it is worthy of emulation.

The Naturalist Sensibility

The extinction of experience refers to more than a reduced exposure to local flora and fauna. It reflects a decline in specific qualities of attention, ways of learning and thinking about the natural world. For most of human history, and until the twentieth century, the absence of a practical knowledge of local flora and fauna seriously jeopardized the material survival of most people. If you lacked this awareness, you didn't eat. The success of industrial expansion temporarily rendered such knowledge archaic, reducing it to the province of specialists and hobbyists, removing it from most forms of schooling, dispensing it to scouting and science. Ironically, this common knowledge gap is exactly what many environmental educators seek to redress, assuming that such intimacy with the natural world cultivates environmental awareness. The naturalist is devoted to such intimate knowledge, studying the woods, fields, and streams to satisfy a deep, insatiable yearning.

What exactly does natural history entail? The two words, joined together, convey nothing less than the history of everything natural, which is (without splitting philosophical hairs) the history of flora and fauna and their environments. This is a very tall order. Indeed, the naturalist's field of study is as deep and wide as the earth, oceans, and sky. Many naturalists narrow their scope of inquiry so as to study a particular species or a specific type of environment. Sadly, the study of natural history is in decline in most university settings. It is eschewed in many departments of biology and even theoretical ecology, simply because such a generalist approach is not in academic favor. Yet most naturalists are generalists. When you observe a bird, you are aware of its habitat, its dietary requirements, its geographic range, its relationship to other species.

The naturalist's first order of attention is the subtleties and nuances of who lives where. Who are the residents of a bioregion? Which species are transient? What are the specific habitats of a particular species? How do various species respond to particular environmental changes? The naturalist will study both the rich textures of microhabitats—looking under logs and in tide pools—and the broader biogeographic patterns—how plant communities vary between hilltops and streambeds. The life history of any animal, plant, or landscape is both a quantitative investigation (subject to collections and experiments) and a narrative experience (a series of stories). The spider's life history (see below) is relevant to scientific ecology *and* it conveys a fascinating biography, replete with adventures and challenges, including perhaps creation myths, and other moral tales. There are scripts and tapestries regarding arrivals and departures—when and where do species come and go? And there are observations of degradation and reinhabitation—how the "fitness" of the environment is modified by its inhabitants. Rhythmic patterns and cycles are the source of endless revelations.

Consider the following passage from Edward O. Wilson's classic work on biodiversity and species extinction, *The Diversity of Life.* In the beautiful opening chapter, Wilson reveals the essence of his naturalist sensibility. He describes how it was "one night at the edge of the rain forest north of Manaus, where I sat in the dark, working my mind through the labyrinths of field biology and ambition, tired, bored, and ready for any chance distraction." He sits in the darkness. He decides to sweep the ground with the beam from his headlamp.

At regular intervals of several meters, intense pinpoints of white light winked on and off with each turning of the lamp. They were reflections from the eyes of wolf spiders, members of the family Lycosidae, on the prowl for insect prey. When spotlighted the spiders froze, allowing me to approach on hands and knees and study them almost at their own level. I could distinguish a wide variety of species by size, color, and hairiness. It struck me how little is known about these creatures of the rain forest, and how deeply satisfying it would be to spend months, years, the rest of my life in this place until I knew all the species by name and every detail of their lives. From specimens beautifully frozen in amber we know that the Lycosidae has survived at least since the beginning of the Oligocene epoch, forty million years ago, and probably much longer. Today a riot of diverse forms occupy the whole world, of which this was only the minutest sample, yet even these species turning about now to watch me from the bare yellow clay could give meaning to the lifetimes of many naturalists.[12]

Wilson combines his acute sensory awareness, his penchant for detailed scientific observation, and his splendid imagination—three interconnected approaches to learning. For Wilson, and an entire generation of remarkable naturalist-writers, these excursions are journeys of self-discovery. The more you study the natural world, the more you understand about the human condition. Natural history is a means of expanding human awareness. Note also how Wilson uses his first-hand observations of wolf spiders to speculate on the broader issue of biodiversity.

Consider some of the qualities of attention contained in this episode. I think they reflect patterns of learning that are synonymous with the naturalist sensibility. The quality that is most worthy of emulation and the deepest source of place-based environmental learning is *the deliberate gaze.*

To gaze is to look intently with curiosity or wonder. To deliberate is to consider what you are viewing in an unhurried, well-considered manner. The deliberate gaze combines wonder, intent, and consideration. This state of mind is crucial for observing the natural world. It requires a kind of improvisational concentration across the senses. Achieving the deliberate gaze requires patience and attention to detail, the ability to stay in one place for an extended period of time—the perseverance to follow tracks in the snow, to peer deeply into the recesses of a tide pool, to listen carefully to the intricacies of birdsong. Contrast the deliberate gaze with the flavor of a video game, when frame rates move so rapidly and you are thrust from one image to another. This is another, perhaps equally important way of seeing the world (and I explore it in much more detail in chapter 6), but for now recognize how the pervasiveness of speed works against the deliberate gaze.

Wilson also conveys great respect for the species he studies. He wishes to collect and systematize all knowledge of wolf spiders (Linnaean legacy), and as he does so, he develops an affiliation with the species. This quality, *interspecies respect,* is also intrinsic to much of the naturalist tradition. In a way, attempting to enter the species space of other creatures, whether this is accomplished through data collection and observation, imagination, sensory awareness, or empathy, takes you out of your own head and into an ecological and evolutionary space. Such a perspective is common in many shamanic hunting rituals. Interspecies respect is best attained and practiced in proximity to the species in the wild. With diminished access to wild places, one wonders how interspecies respect can be nourished.

Wilson's passage suggests other learning pathways typical of the naturalist sensibility. He *collects and systematizes* as a means of organizing his knowledge of the natural world. He strives to ascertain *patterns* in nature. He is a *passionate explorer,* willing to cross the world in his quest to experience the diversity of life. Wilson is the epitome of the reflective naturalist, a person who interprets natural history as a means to better understand the human condition.

I don't assume that all naturalists exude these qualities, that this list encompasses all the different ways that naturalists learn about the world or that being a fine naturalist makes you an honorable person. I merely suggest that these qualities of attention are worthy of emulation as a means for developing a place-based perceptual ecology. The great naturalists have much to teach us in this regard.

Perhaps there is a conceptual correspondence between the loss of biodiversity, the extinction of experience, and the decline of the naturalist sensibility. All of these trends require a revitalization of interest in natural history and scientific ecology. A place-based perceptual ecology is a learning pathway, a means of cultivating environmental awareness, that intends to promote such revitalization. A vibrant learning pathway requires practice and patience, and it is to this project that I now turn.

Practicing (Place-Based) Perceptual Ecology

Any basketball player will tell you that unless you practice your shooting regularly, come game time your jump shot may turn south. Any jazz musician will tell you that you are best prepared to improvise when you practice your scales on a daily basis. Meditation teachers always emphasize the necessity of a daily practice. The same is true for perceptual ecology. Unless you observe the natural world on a daily basis, and develop a practice for achieving this, you are unlikely to cultivate the skills and awareness that perceptual ecology demands.

Count the average number of hours per day you spend inside cars, trains, and airplanes. Add the number of hours you spend watching television and viewing movies. Then add the amount of time you find yourself in front of a computer screen. On top of this, include the hours you spend shopping. Compare this to the amount of time you are observing the natural world. Chances are that your life circumstances don't allow much time for a perceptual ecology practice. The infrastructure of your daily life is filled with too many hindrances and distractions.

Yet there are moments in every day when for one reason or another, you notice the natural world—the cold wind on a snappy winter day as you move between tall buildings in a large city, the colorful sunrise that greets your morning commute, the house finch that sings interminably every spring from his nest outside your office window. These sensory impressions are too noticeable to ignore. Consider how memorable these experiences are and how they stick with you throughout the day.

The remainder of this chapter outlines the basics of a perceptual ecology practice. I describe some simple activities for paying close attention to the immediate landscape, and then explore some ways to approach moving through place and time so as to expand your conceptual reach— all intended as a means for perceiving global environmental change. These approaches require perceptual practice, intended as tools for daily observation. I make this suggestion as an environmental studies professor, with the same zeal that I ask my students to work on a project. But I also place the same demands on myself. I know how easily I'm distracted from a perceptual ecology practice. It takes extraordinary discipline to overcome the momentum of automobiles and computer screens. I also suggest ways of applying this practice to your use of technology (see chapter 6), so as to take advantage of its potential for observing global change—how to place the use of computers and videos in a reflective context, and how to use various forms of transportation to enhance your perceptual ecology practice.

The idea of a practice is that you find a daily means to get more accomplished at something. It requires a *routine and a discipline.* For example, in *Minding the Earth,* Joseph Meeker assumes that everyone has a morning reading, a "morning information session," such as the daily newspaper or the weather forecast.[13] What is the first thing that you think about when you awake in the morning? Meeker suggests that you scan the sky, trees and birds to begin the day, and as a result of such attention, your day will inevitably look and feel very different.

How do you learn about the daily weather? You can watch the weather channel and observe the lows and highs on the map. Or you can take a morning walk with the express intention of reading the day. Your weather forecast will be derived from your observations of cloud formations, the smell of the air, the taste of the moisture, the activity of the birds and insects, the light in the sky. The Weather Channel will provide a context for what you have directly observed. These are the elements of a daily practice.

Good practice requires *mentors and colleagues.* Find other people whom you can share your observations with. Find teachers who have an expertise in making such observations. Often they have finely developed perceptual skills, linked to a solid foundation in natural history and scientific ecology, providing observational methods that you might not discover on your own. Good teachers provide these approaches in a gradual, sequential, cumulative way. There is much good advice in field guides and ecology texts, but there is no substitute for hanging out with a wise, experienced observer.

You can refine a perceptual ecology practice by taking the time to cultivate *personal expression*, either through artistic means, in a teaching venue, or in whatever capacity that will allow you to interpret what you have learned so as to endow your observations with personal meaning. This encourages a deeper assimilation of your sensory impressions, linking them to broader interpretive concepts, whether they are psychological or ecological, or perhaps an intriguing combination of both.

A good practice also requires *improvisational learning.* The attentive basketball player may work within a system, but is always ready to improvise as the game develops. The jazz musician is led by the collective awareness of the ensemble. You never know what you will encounter in your observations of the natural world. You may expect to observe birds carefully when you are distracted by an unusual cloud formation that seems to demand your attention. Mary Catherine Bateson advises that "learning to savor the vertigo of doing without answers or making shift and making do with fragmentary ones opens up the pleasures of recognizing and playing with pattern, finding coherence within complexity, sharing within multiplicity."[14] Such an attitude encourages *active experimentation,* the willingness to try new approaches, to ask new questions, to challenge cherished assumptions.

Observing What You Observe

Consider what you typically notice when you are in a familiar natural setting. If you're interested in wildflowers (whether you're an inspired amateur or a professional botanist), you'll probably observe the flowering plants very carefully. If you are a birdwatcher (whether you're just learning or you've got a life list to die for) you know how easy it is to get totally absorbed in listening to bird songs. In either case, you might finish your walk, furnishing details about the wildflowers or bird life, without having observed much of anything else.

In some teaching situations, to enhance perceptual ecology practice, I ask each member of a group to make a list of what he or she typically notices when "observing" nature. This is a terrific activity to do in the field, after you take a group on a short excursion. Most people will make a list of five to ten categories of observations—for example, mushrooms, birds, trees, watercourses, soils, and the like. I encourage people to compile lists that include patterns of sensory awareness as well—smells, sounds, moisture, colors, and so on. It's helpful for people to compare their lists and even more interesting to pool the whole list of categories. Typically a group is astonished at the depth and detail of the collective list. This reveals an entire spectrum of observational categories and perceptual approaches.

These lists reflect the specific skills, expertise, and interests of the observer. An accomplished artist might be especially sensitive to patterns of light and color—the dappled autumn forest floor. A musician may provide perceptual insight into the sounds of the setting—the humming of insects, the rush of the water, the rustle of leaves. The geomorphologist knows where the glaciers carved the landscape. The meteorologist reads the clouds for signs of tomorrow's weather. The community ecologist describes the patterns of biogeography. Depending on the sensory channels you are most disposed to explore, the expertise you bring to a situation, the special features of the setting you are observing, and an assortment of other factors, every observer brings different levels of awareness, and notices different qualities in a scene.

At different times in your life, for reasons of geography, culture, personal interest, or developmental learning, your observational predispositions may change. Yet there are also some patterns of observation that are intrinsic to your life, special faculties that you work with and develop over time, that provide you with great insight and pleasure. When I was a very young child, I noticed, with some trepidation, how quickly the sky could darken on a hot summer afternoon. A balmy, relaxed sky could quickly become threatening and dangerous. Since then, and because I am endlessly fascinated by the movements of clouds and weather systems, I have always been a passionate observer of the weather. Wherever I travel, I note the relationship between geography and weather—the light of the sky, the smell of the atmosphere, the landforms, and the microclimates. From this observational faculty, I derive insights regarding biogeography, biodiversity, and other broad patterns of life and landscape.

Naturalists and artists use many interesting approaches to enhance their perceptual skills.[15] Some people immerse themselves in their proximate environment, *plunging into worlds of ordinary experience with uncommon focus.* In large measure, this is what Thoreau did during his Walden Pond experiment.

We must learn to reawaken and keep ourselves awake, not by mechanical aids, but by an infinite expectation of the dawn, which does not forsake us in our soundest sleep. I know of no more encouraging fact than the unquestionable ability of man to elevate his life by a conscious endeavor. It is something to be able to paint a particular picture, or to carve a statue, and so to make a few objects beautiful; but it is far more glorious to carve and paint the very atmosphere and medium through which we look, which morally we can do. To affect the quality of the day, that is the highest of arts. Every man is tasked to make his life, even in its details, worthy of the contemplation of his most elevated and critical hour.[16]

More than a century later, poet Diane Ackerman follows Thoreau's advice by plunging fully into the life of the senses. Her wonderful book, *A Natural History of the Senses,* is also an experiment in perceptual depth, an immersion into the most obvious, ubiquitous, and glorious perceptual pathway, the realm of the five senses.

Nothing is more memorable than a smell. One scent can be unexpected, momentary, and fleeting, yet conjure up a childhood summer beside a lake in the Poconos, when wild blueberry bushes teemed with succulent fruit and the opposite sex was as mysterious as space travel; another, hours of passions on a moonlit beach in Florida, while the night-blooming cereus drenched the air with thick curds of perfume and huge sphinx moths visited the cereus in a loud purr of wings; a third, a family dinner of pot roast, noodle pudding, and sweet potatoes, during a myrtle-mad August in a Midwestern town, when both of one's parents were alive. Smells detonate softly in our memory like poignant land mines, hidden under the weedy mass of many years and experiences. Hit a tripwire of smell, and memories explode all at once. A complex vision leaps out of the undergrowth.[17]

These passages illustrate approaches to perceptual ecology practice, inspired by "taking delight in the ordinary,"[18] and using acute observation, with strategic doses of memory and personal reflection, as a means to deepen awareness. Here is an exquisite meld of artistic vision and natural history.

Another avenue for enhancing perception is *rearranging the focus of your daily observations* so that you use underdeveloped sensory faculties, or change the object of your gaze. For example, if you typically direct your attention to the visual, see what happens when you shift your attention to what R. Murray Shafer describes as the soundscape. "Every natural soundscape has its own unique tones and often these are so original as to constitute soundmarks."[19] Steven Feld describes how the Bosavi people, who live in the tropical rainforest of Papua New Guinea, "hear much that they do not see." They develop acute hearing for locational orientation because "much of the forest is visually hidden, whereas sound cannot be hidden."[20] Feld studies Bosavi sense of place through what he terms an "acoustemology" to explore the bodily unity of environment, senses, and the arts.

If you are an avid birdwatcher, shift your attention for several weeks to the lichen. Suddenly you'll notice lichens everywhere, on rocks and trees. You'll observe how their texture and color changes in relationship to moisture, to northern and southern exposures. You'll notice how many different kinds of lichens there are and the habitats in which they are likely to be found.

These examples of enhancing perception have an important common theme—they are subtle approaches that encourage you to change your routine, but they can all be explored within the place where you live, in reference to what you typically experience. Hence you broaden your interpretive scan, and begin to expand the range of your inquiry, while attending to what's in front of you with deeper awareness. This is an important conceptual foundation for perceiving global environmental change.

The Scope of Observation: Landscape and Life Cycle

Let's return to Thoreau, the wise apprehender.

My vicinity affords many good walks; and though for so many years I have walked almost every day, and sometimes for several days together, I have not yet exhausted them. An absolutely new prospect is a great happiness, and I can still get this any afternoon. Two or three hours walking will carry me to as strange a country as I ever expect to see. A single farmhouse which I had not seen before is sometimes as good as the dominions of the King of Dahomey. There is in fact a sort of harmony discoverable between the capabilities of the landscape within a circle of ten miles' radius, or the limits of an afternoon walk, and the threescore years and ten of human life. It will never become quite familiar to you.[21]

Singularly impressive about this passage is the correspondence between life cycle and landscape. Thoreau suggests that the distance he can cover in one day of walking marks a conceptual boundary. The terrain of the landscape is sufficiently varied so that it will take him a lifetime to fully explore it, and even then its familiarity will remain elusive. Intimacy breeds familiarity *and* mystery. The more closely he inspects the country, the greater his lack of knowledge appears. The longer he lives his life, the deeper the mystery of life will seem. There is an existential tension between the full potential of what he can ever hope to understand and the deep, unknowable mystery of the proximate landscape. This is also the source of great delight, the knowledge that he will never exhaust the novelty of what is close at hand, yet forever elusive.

Thoreau also conveys the importance of scale as a means for interpreting landscape. The ten-mile walk affords him a very reasonable observational scope. He intimates that this "radius" is the perceptual limit of his journeys. There is enough within the boundary to keep him fully occupied. It will take "threescore years and ten" of impressions and experiences to notice the variations implicit in his circumambulatory roaming.

Twenty-five years ago, in 1975, I moved from the New York City area to the Monadnock region of southwest New Hampshire. I have now spent approximately twenty-five years of my life in each place. When I first moved to New Hampshire, I was simultaneously enamored and overwhelmed. Every new season, each new location was a complete mystery to me—the long winters, the rugged terrain, how low the cloud cover appeared, the incredible burst of biomass in late May, the blackfly season—there was so much I had to learn about. When I first moved here, I was attracted to the highest places, for they would afford me the best prospect and overview. From various hilltops I could observe biogeographic patterns and watch the movement of the clouds over the landscape.

Now, twenty-five years later, I have lived one generation on the landscape, enough time to raise a family and develop a reasonable familiarity with the human and ecological community. Yet it has taken me this long to even begin to ascertain some of the patterns of the landscape, to observe its rhythms and sequences. Every few years, in a cyclical pattern, we experience acorn storms when, during windy fall days, the prodigious acorn crop falls crashing on the roof of our house. In those years, the squirrels and chipmunks seem equally prodigious. I observe the comings and goings of winter birds—white-breasted nuthatches, evening grosbeaks, chickadees, and tufted titmice. I notice how woodpeckers are more prevalent after damaging ice storms. I've seen the coming of the coyote. I've witnessed one cycle of gypsy moths. I have gained much gardening experience, learning what grows best, given the soil and light conditions of the oak and maple northern forest. The nuances of the climate are more evident, too. I've observed twenty-five years worth of weather patterns, watching storms come from virtually every direction. Although I still relish a trip to a hilltop, I'm more inclined to follow watercourses now, examining the microclimatic nuances of the landscape.

After twenty-five years of observations, patterns now appear that I could never have understood previously. I suppose it will take a full threescore years and ten to put this in much of a perspective. I have taken to bicycle riding more frequently, as Thoreau's ten-mile radius is more appropriate to the scale of a bicycle. I take Thoreau's observations of life cycle and landscape literally. They circumscribe the boundaries of my place-based perceptual ecology practice.

Yet I don't live in a place-and-time bubble. Even as I might accept the narrow limitations of my perceptual view, I still seek to expand that view, to roam freely beyond my place, so as to better understand it. Inevitably, the more intimately familiar I become with this bioregion, the more questions I have about its past and future, about its relationship to other places—its inextricable interconnectedness with complex spheres of ecological and biogeochemical patterns. The remainder of this chapter explores some learning pathways that encourage this broader view, expanding the scope of place-based inquiry so as to gradually encompass a broader perceptual compass.

Landscapes Move through Time

The first step in putting the "history" in "natural" is to consider the origins of your place by providing a context in time. Tom Wessels, in his book *Reading the Forested Landscape: A Natural History of New England,* accomplishes this masterfully. Based on his extensive experience teaching community ecology and natural history, Wessels suggests a series of techniques for investigating landscape. Using active inquiry, keen observation, and aesthetic appreciation, Wessels's approach can be applied anywhere. The essence of his method is a fine foundation for merging landscape and time.

The crux of Wessels's approach is to focus close attention on what he describes as forest disturbance histories, noticing signs and evidence of how the composition of a forest changes over time. For example, using etchings depicting various forest scenes in central New England, he asks the reader a series of questions designed to interpret the historical and ecological factors which tell the story of that scene. You "read" the landscape for clues to its history, finding evidence in signs such as downed trees, dead snags, basal scars, nurse logs, multiple-trunked trees, or stumps. Suggesting both a keen observational eye, a scientific process of hypotheses and data gathering, and a knowledge of various ecoindicators, you learn how to perceive the archival history of the forest, how to "read" it, gaining a sense of the various changes reflected in its contemporary composition.

Wessels provides specific observational instructions for interpreting the landscape. Look for changes in forest composition or age. Consider whether differences between sites are due to shifts in topography or substrate. Look for evidence of six common forms of forest disturbance such as fire, pasturing, logging, blights, beaver, and blowdowns. For each form, Wessels lists the most common types of evidence. For example, downed trees, pillow-and-cradle topography, and decayed nurse logs indicate blowdowns.

Another naturalist who uses a place-based approach to moving through time is John Hanson Mitchell, who describes his experiences in *Ceremonial Time*. When Mitchell moves to Scratch Flat, just west of suburban Boston, he is determined to learn as much as he can about the square mile surrounding his house. The more closely he observes the landscape, the more he notices the relics of time—evidence of previous habitation, the remnants of glaciers, the spearheads he examines at the local historical society. In allowing his imagination to roam, it becomes easier to locate the past in the present, to the extent that his perception of time is dramatically reconfigured.

Mitchell learns how to travel through time in the immediate environs of his backyard. He uses his imagination to locate environmental settings that inspire temporal reflection. On a cold night in the middle of winter, after a heavy snowfall, he sets out for a large glacial boulder in the middle of a neighbor's hay field. He finds the walk fatiguing as he trudges through deep snow. Close to total exhaustion, cold and still a half mile from home, he notices the "huge nothingness of drifting snow." There are no roads or trails, no signs of other people. He feels that he is "thrust back into the very heart of the glacial reign," bearing witness to the "essence of timelessness," seeing "into the heart of the glacier."[22]

In another setting, Mitchell describes how the drumlin on which he lives might have appeared after the retreat of the glaciers, and how at some point, a wooly mammoth must have "walked across the field or meadow" behind his house. "Sometimes on winter walks in the lingering evening light, I like to think that I can see a group of them standing at the woods' edge packed together tightly to protect themselves from the cold, shifting and swaying, facing me, their alert curious trunks exploring the air for my scent."[23]

To interpret landscape history, Mitchell seeks out other community members who can serve as mentors for his explorations. Among his neighbors is "the Red Cowboy," a man who studied glaciers extensively; Nompenekit, who teaches him, among other things, to think in "Indian time"; Margaret Lacey, an elderly woman who could recall summer nights on Scratch Flat in the 1890s. Ultimately, it is the wisdom of his community that teaches him the meaning of ceremonial time, "in which past, present, and future can all be perceived in a single moment."[24] This perspective—integrating spirit, community, and natural history—leads to a "heightened awareness or perception of the way things must have been."[25] Sharing experiences with members of the contemporary human community enhances one's perception of the recent ecological past.

Both Mitchell and Wessels explore the landscape by observing how it shifts through time. Both encourage processes of active experimentation and improvisational learning. They seamlessly merge their fine natural history observations with sensory awareness, layered with community research and open-mindedness. They observe the landscape patterns that are closest to home, the tangible impacts of forest disturbance and human development, the contrasts in ecological communities.[26] As their immersion deepens, they discover that the boundaries between their psyches and the landscape become increasingly interwoven.

Of great interest is how they experiment with scale as a means to expand perceptual awareness. Stepping between landscapes and time frames provides a means of comparison and perspective. Here lies another important conceptual challenge for perceiving global environmental change—the necessity of exploring scale. The more immersed you become in the local, the more necessary it becomes to juxtapose place and time. Scale is a conceptual language that allows for the amplification of perception. It's a tool for detecting patterns of connection across the boundaries of ecology, geography, history, and psyche.

Expanding Your Umwelt

Jacob von Uexkull, a philosophical anthropologist, conceived the term *umwelt* in 1920 to refer to the physiological perceptual environment of any biological organism. He recognized that every organism has a perceptual potential determined both by its organismic qualities and how it responds to environmental stimuli. Umwelt refers to the capabilities and limitations of the senses. How far can you see? What do you hear? What scents do you smell? The dragonfly, with its kaleidoscopic vision, or the bee, with its sensitivity to ultraviolet, "sees" the world very differently than you and I. Different critters experience different sensory environments. And different individuals perceive the world variably depending on a matrix of environmental conditions. If you have "normal" vision, compare your view of the world with someone who is colorblind, or nearsighted. Similarly, your view of the world will be very different if you grew up in the Arctic, or if you are from the tropics.[27]

It turns out that the idea of umwelt is also of great ecological relevance. Other organisms dwell in perceptual environments that are completely unnoticed by humans. A tick may lay in waiting for years before its food source comes along. Ecologists Timothy Allen and Thomas Hoekstra observe that the size of the umwelt of an organism, in relationship to the human umwelt, determines how organisms are selected for ecological research.[28] For example, birds view a wider landscape than large mammals. They are useful for island biogeography studies because they give us a great deal of information regarding the movement of species between places. Allen and Hoekstra suggest that the idea of umwelt is of great conceptual power in placing organisms within a larger ecological framework, because it is a tangible, physiological means through which different perceptual scales can be explored and compared. This is one research pathway within what ecologists call "variable scalar hierarchies."

The depth of the relevance of umwelt for ecological research is beyond our immediate interest, but what I find intriguing for perceptual ecology is the idea that humans have the capacity to expand their umwelt, so as to explore spatial and temporal scales well beyond their organismic limitations. This has special relevance to studying global environmental change. To interpret global patterns, you must move beyond the limitations of "common sense," or that which you observe on a daily basis, even what you perceive over a lifetime, in order to consider a full range of spatial and temporal possibilities. What did the place where you live look like ten thousand years ago? What will it be like in the future? John Mitchell and Tom Wessels merge imagination and detailed observations of landscape ecology to move beyond what is immediately apparent. This allows them to conceptualize paleoecological landscapes. They are engaged in the expansion of umwelt.

As David Rothenberg insightfully writes, technology represents the amplification of umwelt, changing how we perceive nature, but also reminding us that it "always remains beyond us, something to wish for, still far away, just out of reach."[29] By looking through a telescope you change the scale of your vision, making distant objects seem close, shrinking the expanse of time and space. By looking through a microscope, you get much closer to the minuscule, noticing how much space there is in even the smallest distances. How can the expansion and amplification of umwelt open the doors of perceptual ecology while enhancing the stability and groundedness of place-based observations? Let's consider this challenge in more depth by experimenting with the juxtaposition of scale, starting from a series of local observations.

I tend a small garden plot, embedded in the oak and maple hardwoods of southwest New Hampshire. I have worked the soil for twenty years, cultivating this tidy portion of forest landscape. If I leave the plot untended for any length of time it is quickly reclaimed by the successional tempo of the forest, starting with the virulent, thorny blackberries and the ubiquitous oak seedlings sprouting from last year's acorns. My garden is merely a short-term modification of a successional forest pattern that transcends my limited efforts. As I work the soil I plunge my spade into countless pebbles and small boulders. To understand where they come from I must step out of my organismic umwelt and incorporate the conceptualization of a larger time frame. These rocks were deposited during the last glacial interval when the landscape was covered with ice, not so many years ago in geological terms (about 12,000 years), but many more years than I can readily perceive. To gain a full context for understanding why the soil is rocky, I have to step out of my immediate place and time (in this case, with the help of books) to understand (and imagine) how the landscape was covered with boulders and pebbles.

I notice the edges of the garden, framed by stone walls constructed from the glacial till. The wall marks an edge, separating the garden from the broad expanse of the northern forest. The edge represents a unique habitat, attracting a variety of birds and mammals that otherwise may not choose to live here. Each of these creatures has a sense-specific umwelt to which it responds. When birds choose to eat berries from the thorny blackberry bush, we share a food source, and to a certain extent there is a resonance in our umwelt. We both perceive the blackberries as food in the same place and time. This becomes much clearer when the woodchuck invades the kohlrabi and we compete for the same food source. Also, there are thousands of organisms that I never directly encounter—insects and soil microorganisms—whose role in this small garden patch is crucial, but whose umwelts share little tangible resonance with my own. Can I use my imagination to explore the sensory worlds of the birds, mammals, insects, and soil microorganisms, so as to gather a more integrated awareness of the composite, ecological umwelt of this edge in the forest?

An interesting way to expand umwelt is to *shift between the large and the small*. Try following an ant as it moves across the forest floor, or as it zips across your kitchen. See if you can spend fifteen minutes, or even longer, observing any microhabitat—the lichens on a boulder or tree trunk, soil microorganisms, pond scum. When you are fully immersed in the microworld, shift your attention to the sky. Follow a cloud from its formation to its dissipation. Watch a hawk as it soars through the landscape. Observe an airplane as it crosses the horizon. These shifts in perception allow you to expand or contract your gaze accordingly, permitting you to play with the boundaries of perception. They are a means to expand umwelt without technological aids and are a good foundation for shifting perceptual boundaries.

These series of observations reflect the necessity of using scale to interpret the patterns of place. I must expand my sense of time to understand why there are so many boulders in the garden. I must shift my umwelt to even begin to grasp how other creatures use this same garden space and the edge thereby created. Both processes involve forms of perceptual reciprocity. I must be open to the proximate sensory world in order to spur my imagination to expanded temporal and organismic perspectives.

Now I'd like to suggest how local observations may lead to questions about other places, prompting shifts in spatial scale. This next example depicts how local ecological knowledge leads to questions about global environmental change. For years, my neighbor has been participating in spring migration bird counts. Originally an amateur birder, after years of patient observation and study, she is now highly skilled at bird identification. She has gained this knowledge by spending hours in the field, poring through field guides, and by hanging out with more experienced ornithologists. She is a member of a community of bird observers, people who share their common interests and concerns. Through this growing interest and expertise, she develops a stunning, acute sense of hearing and vision. She's able to locate nondescript sparrows in thick brush. She is immersed in her observations of birds and their habitats, noticing what various birds eat, where they nest, whether and when they migrate.

One year the community of bird observers recognizes a distinct pattern. A particular species is declining, and soon, it appears, it will significantly shift its range, or, perhaps, its survival will be threatened. To fully understand the ecological consequences of these patterns, several juxtapositions of scale are required. What are the migration patterns of these birds? What are the conditions like in their winter residences? In their summer breeding territories? Is the species declining because of rainforest disruptions in Latin America? Or because of logging in the northern forest? Or because it's getting too warm in New England? These are sound ecological questions that emerge from her place-based observations, but they require information from other places. The disappearance of this bird may have ramifications for the whole forest. It may be a sign of forest decline, of shifting habitats and ecosystems. It has long-range ecological implications that can only be ascertained if local naturalists from different places share their information. For my neighbor, a seasoned natural history observer, global environmental change is no longer a vague premonition. Her local ecological knowledge provides a place-based perceptual groundwork for her concerns. All of her good questions require a shift in spatial scale.

None of these observations (from the garden plot to the annual bird count) would be possible without three closely connected learning pathways—paying careful attention to local natural history, the ability to expand and amplify organismic umwelt, and the ability to experiment (perceptually) with scale. Timothy Allen and Thomas Hoekstra in *Toward a Unified Ecology* construct an ecological epistemology emphasizing the juxtaposition of scale as a means of interpreting complex patterns. Their premises are that all ecological processes are multiscaled and those processes that best match human scales of unaided perception are the ones we know best. Ecological interpretation requires the use of multiple criteria—landscapes, ecosystems, communities, organisms, populations, biomes, and biospheres. The intelligent observer will use several of these criteria (or scales) depending on the specific relationships of interest. Note their guidelines regarding the *juxtaposition of scale*: "For any level of aggregation, it is necessary to look both to larger scales to understand the context and to smaller scales to understand mechanism; anything else would be incomplete. For an adequate understanding leading to robust prediction it is necessary to consider three levels at once: (1) the level in question; (2) the level below that gives mechanisms; and (3) the level above that gives context, role, or significance."[30]

This commonsense observation has extraordinary theoretical and practical power, yet it encompasses a perceptual wisdom that is often neglected and easily forgotten. It provides wonderful guidance to how a place-based perceptual ecology can lead to the interpretation of global environmental change.

For example, I live in the dry oak-beech-maple, eastern deciduous northern woodlands. Through familiarity and study I recognize the most prevalent flora and fauna of the habitat, although I have virtually no knowledge of the soil microorganisms (a scale shift). To thoroughly interpret the "level in question" I should be able to identify, study, and observe most of the plants that live in a chosen plot of woodland. This allows me to notice subtle changes and nuances, especially those that occur within measurable and perceivable time frames.

If I observe drought conditions—a lack of rainfall, water-starved plants, or diminished bioproductivity, my insights are incomplete without linking these observations to additional interpretive levels. Why is there no rain? To answer this, I observe atmospheric circulation. This requires reviewing weather maps of the present and past, to ascertain how this drought pattern corresponds to both global weather phenomena and longer-term climatic data. Or perhaps I must rely on the deepest levels of intuition and awareness, finding patterns in the movement of the clouds, in the wind direction, and changes in how the air tastes, smells, and even weighs. And how do plants and animals respond to this lack of rainfall? What information do I need to help me understand the cellular level of response? Why this drought, when will it end, and what measures must our community take?[31]

My interpretive power is limited to specific time and space boundaries which are constrained by two qualities: my perceptual limits (what I can see or measure) and my conceptual limits (the extent to which I incorporate multiple levels in my thinking). From a scientific perspective, my understanding of mechanism will be greatly enhanced by attending to microscopic scales, which may include soil chemistry, microbial biology, plant physiography, and other sets of inquiry. And my understanding of context requires attention to biogeography, paleoecology, climatology, and geomorphology. I don't want to limit these categories to the conceptual frameworks of contemporary science. As David Abram suggests, this inquiry penetrates shamanic territory as well—investigating underworlds and overworlds—attending to the boundaries between worlds, for it is at the boundaries that you gain the most intriguing insights.[32] *Extraordinary insights occur when we perceive the relationships across and between boundaries.*

Moving between Places

Such boundary shifts often appear more striking while traveling, when you move between places. Earlier in the chapter I suggested that it may take at least one generation of living in a place before certain observations begin to make sense. But consider the reverse sequence. Visit a place that you once knew well but haven't seen in quite some time. Depending on how long you've been away, chances are that you'll be very impressed with the magnitude of the changes that are revealed after some years have passed.

During the summer of 1968, I worked on an American Friends Service Project in Union City, California. At that time, Union City was shifting from a rural to a suburban community, but there was still plenty of open space and relatively undeveloped landscape. South of Union City, toward San Jose, the landscape was mainly rural. Twenty-five years later (one generation) I was in the Bay Area and I had occasion to visit Union City. It was unrecognizable. The land from Hayward to San Jose was completely developed. I couldn't even locate the street where I had lived, as new housing developments obliterated the scenes so deeply etched in my memory.

When you become intimately familiar with a place, it can be hard to interpret the cumulative changes. But separated by distance and time, these same changes may be shockingly evident. I realized that what happened in Union City was representative of an urbanization process that was taking place all over California. Indeed, it resembled patterns of land use development for much of North America and much of the world. It was direct, tangible evidence of how rapid economic development contributes to global environmental change. I witnessed a transformed ecological landscape. Many people who are concerned about environmental issues develop their concern when they return to a childhood place and see how it has been thoroughly overhauled. As John Elder asks, "where can the wholeness of life be found when the hills of childhood are subdivided?"[33]

In a way, returning to a place where you once lived allows you to *move backward and forward in time.* Your memories represent the past in contrast to what you observe in the present, allowing you to feel like you're observing the future. When I returned to Union City, I felt like I had moved into the future, although I was just being brought up to date. The perceptual challenge is to engage in such travels by interweaving memories of the past with sensory observations in the present. In this way, traveling into your past is less a nostalgic exercise and more an opportunity to observe global environmental change. It allows you to take a measurable period of abstract time with all of the graphs, charts, and statistics that otherwise describe it and experience it more directly. No land use atlas could have provided me with the visceral understanding I gained by revisiting Union City.

Moving between places also allows you to return to your home place with an entirely new perspective. During a recent trip to the Olympic peninsula, I was immersed in life zones, temperate rainforests, biogeographic patterns, weather patterns, olfactory sensations, and microclimatic variations that were dramatically different from my everyday, eastern deciduous woodland experiences. During a backpacking trip I was deeply engaged in the sensory world, practicing perceptual ecology so I could learn as much as I could from this extraordinary place. Upon returning home, when I surveyed the forest, those trees that previously seemed so large now appeared as spindly weeds. Their majesty seemed ruthlessly challenged. Everything appeared so much smaller! The trip to the Olympics allowed me to attain "beginner's eyes" in perceiving my home place. Similarly, when returning from the desert Southwest, I gained a new appreciation for humidity, never before realizing how much I take moisture for granted. I'm not sure I would have had these impressions without developing a place-based perceptual ecology practice. My familiarity with a home place gave me a context for comparison.

You don't need to travel across the continent to move between places. Going from urban to rural areas, or vice versa, provides interesting lessons in density and speed (more about this in chapter 6). Moving from land to sea, from plains to hills, from field to forest—anytime you move between places you have an opportunity to enhance perception. By doing so, you can compare places readily. What becomes increasingly apparent, as you compare places, is the prevalence of edges, the boundaries that mark differences. Exploring the edge is a very powerful way to juxtapose scale, as the edge makes you aware of how the shifting of boundaries is also a function of how you perceive something.

Exploring the Edge

In *Seeing Nature,* Paul Krafel describes how observing edges is the key to interpreting environmental change. His wonderful little handbook suggests dozens of perceptual experiments that enhance ecological awareness through direct field observation. He devises ingenious ways of marking shifting boundaries—placing pine cones on the edge of a melting snowbank, placing sticks on a beach to observe the shifting tide, surrounding plants with clumps of stones to determine whether they are expanding their range. [34]

These techniques alert Krafel to changes that otherwise might go unobserved. Some of these observations merely mark local rhythms and cycles, but others track the relationship between places—the migration of species or changing climatic phenomena. Observing such changes allows Krafel to interpret both spatial and temporal edges. He suggests that you can move your eyes back and forth through time by gazing upstream and downstream. For example, by looking downstream he notices leaves rotting into brown slime, whereas upstream the brown slime freshens into leaves. He watches flowing water as it moves through the landscape and by doing so, he can glimpse ahead or backward to witness the past and future of this portion of the stream. "Moving back and forth helps me sense how things once were and how things will be in the future."[35]

These types of observations can be made locally, yet they alert you to broader global patterns. It's late January in the hill country of southwest New Hampshire, yet it's raining and there are robins singing. Why are various bird species moving northward? There are many ways I can observe shifting edges in my home place. I notice that the migrating warblers seem to arrive a little bit earlier each year. It seems that mosquitoes are here later each autumn. I notice how the rain-snow line on the weather map is slowly inching northward. I read reports which suggest that the tundra is retreating in Alaska, that glaciers are diminishing in many alpine environments. All of these observations share a common trait—they are environmental changes that you can visibly track, by carefully noting the movement of a boundary. Edges subtly shift and shimmer.

Observing edges requires perceptual flexibility, the ability to use a boundary as a means for integrating and separating diverse landscapes and habitats. How boundaries change, and the various means of marking edges is crucial to landscape ecology, and is at the core of much research in biodiversity studies. Interestingly, this approach is now incorporated into a new generation of field guides. Particularly impressive (along with Wessels's approach cited above) are John Kricher's two guides to eastern and western forests in the Peterson series.[36] These books emphasize indicator species as a means for recognizing various habitats, and they focus on successional changes, disturbances, and adaptations—ways of observing edges.

Krafel alludes to another technique that requires further elaboration—*marking time.* When you mark an edge, noting how it shifts boundaries according to various climatic, topographic, seasonal, or ecological rhythms and cycles, you are able to track how that boundary moves through time. Wessels and Mitchell are engaged in a similar process. By tracking forest disturbance histories, or searching for paleoecological remnants, they look for signs of environmental change in the landscape. In a way, when you observe how edges shift, either by marking them with tangible signs, or by imagining how the edge might move, you are engaged in a process of *making the invisible visible.* This approach is also fundamental to perceiving global environmental change. Environmental actions have impacts across spatial and temporal boundaries that may not be immediately apparent. How can perceptual ecology make the invisible visible? We return to this challenge in chapter 5.

Place-Based Perceptual Ecology and Beyond

The premise of this chapter is that as you develop a perceptual ecology practice, and as your local natural history observations become increasingly acute, you develop your ability to perceive environmental change. These observations become an aspect of everyday awareness, prompted by the signs of seemingly ordinary events, but inspired by serendipity, immanence, discontinuity, and mindfulness. It takes diligence and perseverance to understand how to place your observations in perspective. For example, the ability to expand umwelt, or to juxtapose scale, requires perceptual alacrity and flexibility. These are qualities that must be practiced, taught, and learned. Much of this chapter provides suggestions and examples, merely a small sample of what is possible, of how place-based perceptual ecology may be approached. Dozens of variations will emerge from different learning environments—depending on where you live, how old you are, how much experience you have, your observational inclinations, and how much natural history and scientific ecology you've been exposed to.

Most of the examples in this chapter are limited to relatively tangible observations and comparisons. Chapter 5, "Interpreting the Biosphere" uses these approaches as a means to move through broader expanses of geographic space and geological time. Chapter 6, "The Internet, the Interstate, and the Biosphere," considers the challenge of advanced technologies of speed, distance, and information and how these remarkable devices allow for extraordinary perceptual possibilities while they simultaneously threaten a place-based perspective.

Our challenge is to explore how place-based perceptual ecology is a gateway to a broader conceptual expanse—understanding biogeochemical cycles, Gaian processes, biodiversity studies, the history of life on earth—all prerequisites of interpreting global environmental change.

This requires the perceptual flexibility of moving between multiple perceptual worlds, investigating deep time, and traversing scales. These are the conceptual foundations of a biospheric perspective. How do we make place-based perceptual observations so tangible, so visceral, so meaningful, that they resonate from the biosphere to the backyard, integrating local and global phenomena so that we truly bring the biosphere home?

We are on the verge of an eclectic, unprecedented environmental science. This extraordinary synthesis transcends disciplines, connects local and global ecologies, observes the flow of biogeochemical cycles, and interprets the evolution of life on earth. These are the prerequisites of twenty-first century global change science, a way of knowing that stretches the mind like an accordion, moving from bacteria to biosphere through the interface of organism and the environment. Dwelling in place, but searching the pre-Cambrian past to discover roots and origins. Moving through space while scanning the biosphere to understand the flow of matter and energy. To even glimpse the complexity and grandeur of global change takes you beyond ecology and evolution into the boundaries of perception—a biospheric perspective, if you will, that shakes the very foundations of what it means to be human.

Attaining this perspective is no simple matter. It is not immediately obvious that the earth is 4.5 billion years old, or that North America has been covered by glaciers many times over, or that microbial organisms can alter the composition of the atmosphere. It's hard enough to uncover the layers of ecological relationships in the place where you live. How can your daily natural history observations be expanded so as to encompass a biospheric perspective?

This chapter explores the learning pathways that make such a perspective more prevalent and accessible. I superimpose a biospheric gaze on place-based environmental learning, providing some perceptual bridges that allow you to span the intimacy of place and the layers of the biosphere. All of my intuition and experience as an environmental educator suggests that the broadest conceptual breakthroughs are most likely to occur when a person learns how to reconceive the familiar. The reconceptualization of space and time begins in the here and now. What if you learned how to breathe as if every respiration was an act of biogeochemical recycling? Or if you learned how to look at an arthropod as a long-lost Ordovician cousin? What if you looked at clouds as the interface of air and water, tracks of moisture in a complex global climatic system? These are venues of everyday awareness, habits of perception that build learning.

I begin with a perceptual approach inspired by Vladimir Vernadsky, the world's first biospheric scientist and author of *The Biosphere.* Vernadsky suggested that with patience and perseverance you can observe the dynamic layers of life in the biosphere. There is much more to a landscape than what you initially perceive. Vernadsky was an exemplary biospheric naturalist, a person whose local natural history observations incorporated an understanding of global ecological processes. I describe how biospheric naturalists embody special perceptual qualities. They share an extraordinary ability to integrate analysis, imagination, and compassion—the three pillars of place-based perceptual ecology. In addition, their ability to interpret the biosphere is grounded in their facility with biospheric knowledge systems—familiarity with biogeochemical cycles, the geological time scale, and the five kingdoms of planetary life.

How might these approaches become relevant to a daily perceptual ecology practice? How do we use them to interpret the biosphere and to track environmental change? In the second half of the chapter, I consider how such basic observations as following a storm system or studying a garden can incorporate biospheric natural history. Observing the weather is a means to consider global circulatory systems as they move through and between places. It's a way to track the four elements—fire, air, earth, and water—as indicators of environmental change. Observing what lives in your garden allows you to attain a five kingdoms perspective (bacteria, protists, fungi, animals, and plants). You learn about the biosphere by becoming familiar with the full spectrum of evolutionary possibility. You gain a better sense of human lineage and ancestry, lending clarity to concepts such as biodiversity and megaextinction.

Interpreting the biosphere is such a profound conceptual challenge because it entails stunning juxtapositions of scale—moving from a fifteen-minute thunderstorm to a million-year climatological trend, shifting from tending your garden soil to observing the patterns and trends of biodiversity. In making these shifts, in learning how to move through diverse spatial and temporal realms, you are more likely to perceive global environmental change. It all starts with how you choose to observe your proximate environment, the questions you ask, the linkages you make, the knowledge that you have access to, and the extent to which you experiment with perception.

Vernadsky's Perceptual Challenge

Although this movement is continually taking place around us, we hardly notice it, grasping only the general result that nature offers us—the beauty and diversity of form, color, and movement. We view the fields and forests with their flora and fauna, and the lakes, seas, and soil with their abundance of life, as though the movement did not exist. We see the static result of the dynamic equilibrium of these movements, but only rarely can we observe them directly.[1]

Perched on a cliff I peer at the open expanse of the Gulf of Maine. I sit at the edge of a small island ten miles from shore. High pressure has been building for several days. The air is dry. The sky is cloudless and bright blue. There's a light, southwest breeze. This is a midsummer weather pattern in the middle of May. Imagine Mediterranean Maine.

What an expansive panorama. Land, sky, and sea meet at sharply defined boundaries. I see the low hills of the coastline in the distance, a thin layer of land at the edge of the Gulf of Maine, sandwiched between the dark-blue water and the light-blue sky. Four elements—a sliver of earth, bounded by infinite water and air, with the fire of the sun warming my body.

At first impression I am soothed by the seeming tranquillity of this scene. Yet this is a prospect of subtle and dramatic movement. The gentle wind pushes the sea. The whole ocean surface is undulating, a continuous rolling motion. You can watch the black-and-white eider ducks bobbing on the blue waves. The same breeze that cools my body creates a ripple effect on the water. Undulations and ripples emanate in endless motion, penetrating my skin just as they pass through the water and air.

I notice a thin, translucent cirrus cloud, a sign of moisture in the upper regions of the atmosphere. Clouds allow you to track the path of water as it moves through the sky. Lowering my gaze, I spot the tide line against the rocky shore. Here the barnacles and seaweed meet the lichen-covered rocks. The swirling tide covers and uncovers the coastline. Watching the tide allows you to follow the dynamic changes at the edge of the sea. Tide pools are hidden and then revealed. The changing color of the lichen tracks the spray of moisture. The tide is also a constant reminder of the orbiting moon.

A migrating chestnut-sided warbler lands on the small spruce tree just a few feet in the distance. Warblers rest and feed by day and travel by night during their long trek from the subtropics to the northern forests. When the wind direction is southwesterly, and weather fronts are properly aligned, warblers pass through here in waves. This island is a fine resting point in their migrating northward flow. Observing this flow allows you to track the movement of spring.

One of the reasons why I am so attracted to this spot is because I can witness the interface of air, sea, and land. I can observe currents of change by watching how the wind and water interact, how they sweep across the land and carry life in their grasp. Later on, at sunset, or on a cloudy day, or perhaps in the midst of a turbulent storm, the vista will be transformed—it may appear mysterious, murky, chaotic, or unsettled. The colors of ocean and sky shift throughout the day and night, sometimes subtly, at other times dramatically. When the sun is higher, the light will dapple the ripples, forming glistening sparkles, a shimmering effect similar to an impressionist painting.

I recall a squall I once witnessed here. The cloud line rolled forward, a foreboding gray-black mass, swallowing all in its wake, covering the scene like a sentient blanket, a message from Mordor. How eerily silent it was in its movement, yet inexorable and relentless. So often fog obscures this place, turning the sky grayish-white and the ocean silvery-green. I remember a driving northeaster that lasted for eight days—a slough of lows lined up along the New England coast as if on an assembly line waiting their turn to unleash torrents of rain and wind. Moisture poured from the dark-gray sky, with sheets of rain catapulted by the wind, moving horizontally so that the ocean and sky were in furious tandem. The paths and trails of the island became rivulets and puddles.

From this spot I can summon a melange of memories. The dates of the events no longer matter. Rather I experience layers of images of four elements at play, intermingling through the swirling biosphere. Views and scenes I've observed roll together, indistinguishable in my timeless reverie.

It is through the meshing of stillness and memory that I observe the flows of this spot, that I learn to perceive the biosphere. It's a wonderful peculiarity of mind that the more still you become, the more prevalent motion appears. Picture the landscape when you fly in an airplane or when you are zipping at high speeds along the interstate. It just seems to pass by as if static, like the backdrop of a Hollywood movie set. The faster you move, whether it's via high-speed transportation, or through a whirling, speedy mind, the more the external world appears as a fixed scene. But when you retreat into stillness and your mind slows down—

calming, steady, open—the longer you allow yourself to sit and observe, the more dynamic the external world appears. You can feel the slightest motion in the air through your body, or in the vibration of sound. The air itself seems to vibrate, molecule by molecule, as if you could hitch a ride on a floating atom.

Gaston Bachelard observed that "memories are motionless and the more securely they are fixed in space, the sounder they are."[2] Hence the irrevocable attachment of memory and place. I return to this island spot every May to observe the four elements in motion, and to dwell not only in the place as it is now but to reside in the memories of the countless times I have been here before. Stillness yields memory. It is through my rootedness to this spot, the stillness of memory, that my body and psyche merge with the scene so that sky, sun, land, and water are a collage of movements in the passage of time. I aspire to sail on the vapor of memory as the fog rises from the ocean and lifts to the sky.

I have an intuition of biospheric perception.[3] Surely the movement and energy here is intricately complex, inconceivably variable, replete with cycles and patterns, more involved than I can ascertain. Vernadsky suggests that the biosphere contains layers and levels of movement, most of which are imperceptible, hence his implied challenge—how to grasp the subtleties of changes that may otherwise pass you by.

In his classic work, *The Biosphere,* in a section on "The Multiplication of Organisms and Geochemical Energy in Living Matter," Vernadsky asserts the ubiquity of life. The energy of life unifies the biosphere. "The diffusion of living matter on the planet's surface is an inevitable movement caused by new organisms, which derive from multiplication and occupy new places in the biosphere; this diffusion is the autonomous energy of life in the biosphere, and becomes known through the transformation of chemical elements and the creation of new matter from them. We shall call this energy the geochemical energy of life in the biosphere."[4]

Here, from my island viewpoint, there is much more that I can observe if my sensory impressions are also informed by knowledge and study. The transformations and movements described by Vernadsky require depths of scale and perception that transcend my unaided eye. What else might I learn about this place if the latest scientific research and theoretical interpretations inform me?

Without biological oceanography, how would you know that there are cycles of phytoplankton productivity that correspond to the availability of nutrients, and that these cycles contribute to the mixture of air and sea currents, prompting convective overturns and eutrophic layers? These cycles are crucial for the health of the fishery and the seabird populations. Without earth systems science, how would you know that nutrient cycling is connected to a complex network of photosynthesis and biogeochemical cycles? Without geology and geomorphology, how would you know that this island is a high point in a drowned coastline, formed by the melting of the ice and the rise of the sea? All of this too is happening right before my eyes, although I would never know these things without the diligent observations of several generations of scientists.[5] The limited framework of my observational capabilities is the tiniest opening in the window of biospheric perception.

What is Vernadsky really getting at? He suggests that behind, beneath, and between the scenes, at a level imperceptible to the unaided eye, there are microbial layers of activity, biological and chemical transformations that are crucial to the metabolic and physiological processes of the entire planet. "The creative wave of organic matter" is the very foundation of the biosphere. Life is a powerful geological force. The ceaseless motion of this coastal scene is more than the infinite combinations of the four elements. It embodies the activity of living matter, a precise coordination—the simultaneous breathing of life and atmosphere, the cycling of carbon between earth, rocks, ocean, and air. Life is minerals and minerals are life.

Vernadsky speculates on the ubiquity of this activity and in so doing he provides a scientist's corollary to the Buddhist notion of samsara, the endless wheel of life and death. "It is hard for the mind to grasp the colossal amounts of living matter that are created, and that decompose each day in a vast, dynamic equilibrium of death, birth, metabolism, and growth. Who can calculate the number of individuals continually being born and dying?"[6]

Here is the essence of Vernadsky's perceptual challenge. From your observational outpost, whether it's an island perch, or a field of wildflowers or a small garden plot in the middle of the city, can you observe your place as a node in the biosphere—a matrix of life and death, microbial activity, energy exchange, and biogeochemical transformation? Can you sit still long enough to let stillness yield memory, and in so doing, allow the levels and layers of movement to permeate your awareness? And then finally, when your observational skills are receptive and tuned, perhaps you will see yourself as a body in the biosphere, a reflective mind in the mirror of creation.

Cultivating biospheric perception requires close attention to the ubiquity of life, its continuous movement through space, and its forms evolving through time. Such movement and evolution occur at spatial and temporal scales that are beyond direct observational possibilities. It's through experimentation at the microbial and molecular level that we ascertain biogeochemical cycles. It takes fossil evidence to uncover the history of life in conjunction with the movement of tectonic plates. Scientists use an array of geochemical, physical, and biological evidence, bolstered by isotopic and nonisotopic dating techniques, to determine rates of environmental change. Computer programs are used to develop biogeochemical and climatic models that simulate the movement of energy in the biosphere. Vernadsky compels us to interpret these data from a biospheric perspective.

Just as Darwin's legacy of evolutionary biology allowed for the reconceptualization of the origins and development of life, spawning the sciences of ecology and biodiversity, so does Vernadsky's legacy of earth systems science inspire the reconceptualization of the earth. Both legacies imply perceptual demands that are barely incorporated in most educational settings. Although evolution is widely taught, few people can provide the rich details of the history of life on earth, or have the geological time scale etched in their memory. Similarly, biogeochemical cycles are hardly a well-known concept, and the idea of the biosphere is relegated to short sections of environmental science texts. How can global environmental change be interpreted without these fundamental concepts in hand? Vernadsky's perceptual challenge is an educational imperative—how to look at the world from a biospheric perspective.[7]

This is a tall order for place-based environmental learning, an approach that relies on tangible observations of what is close at hand. The biosphere and its rich tapestry of life may surround you at all times, but it is not so easy to grasp. It is intangible, conceptual, and abstract. So what does it take to pry open the doors of biospheric perception? Throughout this chapter, I provide some examples of how intimacy and familiarity with place, the close study of what is proximate, leads the observer to wider scalar realms. I have found that the more deeply I explore my island perch, the wider my gaze becomes. But this only occurs if my explorations are supported by good science, imaginative metaphor, and compassionate identification. In the next section, I explain why this is so, and how it is borne out in the work of exemplary biospheric naturalists.

Exemplary Biospheric Naturalists

In his delightful book *Life: A Natural History of the First Four Billion Years of Life on Earth*, Richard Fortey, a paleontologist at the Natural History Museum in London, describes his first scientific expedition. Searching for ancient fossils near Spitsbergen, Norway, he is utterly enamored with the "shimmering textbook of zoology" that surrounds him. "There seemed nothing to interfere with the joy of observation, no end to knowledge, no possibility that any discovery should be less than astounding."[8]

For several months, Fortey collects fossils along the inhospitable, damp, chilled, but exhilarating glacial coastline, unearthing hundreds of fossil specimens from the limestone bedrock. He collects Ordovician trilobites and graptolites (an extinct planktonic animal), fossils that are nearly 480 million years old. Fortey moves alternately from collecting site to tent as the squalls, driving drizzle, and summer snow of this Arctic coastline often prompt him to seek shelter. From the "feathered haven" of his eiderdown sleeping bag he listens to the sea pound against the shore, noting that "the waves that broke on the cold, gray Arctic stones would have sounded the same in the Ordovician or the Jurassic."[9]

Twenty years later, as Fortey reflects on the various expeditions that inform his paleontology, he admits that his Ordovician journey is more than a scientific odyssey, but a personal vision quest as well.

I have spent much of the last twenty years trying to imagine what it would be like to wallow in Ordovician oceans; on my own journey through geological time this is where I have lingered most in my own halting voyage to enlightenment. It is impossible to catch more than glimpses of such long-vanished worlds. The excitement of discovery comes . . . when clarity breaks through the dimming confusion of time to vouchsafe a sight of something hitherto obscure, much as the summer sun dispels a morning mist to disclose an unexpected and pleasing landscape. In truth, the past is never completely revealed, for imperfections in the record of the rocks guarantee that parts of the view will remain indistinct. Nor can a few inquisitive souls reveal more than a handful of scenes from an Earth that was already complex and varied 450 million years ago. The constancy of ecology and marine habitats provides no more than crude shafts of light into the murky past. None the less, the endurance of these habitats and their varied ecological opportunities has provided at least a few familiar reference points, while so many other aspects of history have been more fickle and uncertain. Mutability has ruled: the face of the world has transformed many times, sea-changed again into something rich and strange.[10]

Throughout *Life*, Fortey uses his imagination to identify with these rich and strange habitats. He imagines himself standing on a Cambrian shore in the evening, searching through heaps of seaweed for trilobites and arthropods. Observing pre-Cambrian stromatolites, he envisions an "astoundingly ancient ecology,"[11] marveling at how a community of microbes can endure for three thousand million years. While reconstructing fossil evidence, Fortey visits the Devonian age and speculates on the origins of liverworts, mosses, clubmosses, and ferns, imagining the era of the "great greening."[12] He takes a walk through a Carboniferous coal forest, a 330-million-year-old scene, replete with trees and hidden creatures. "The only sounds are generated by the scrape of insect limbs, or maybe a low amphibious hiss. There is little color, because there are no plants that have yet developed the flamboyance of flowers."[13]

What audacity it takes to write a one-volume, four-billion-year history of life on earth. How can such a history possibly be written? Fortey is compelled to attempt such a project because he wants to understand the context and history of his humanity. With every fossil discovery the context becomes more complete. At each geological era, with the unfolding of another ancient earth vista, Fortey writes another chapter of his origins story. The story of life is literally unearthed through fossil evidence until finally he attains a glimpse of deep time. The ancestry and lineage of humanity develops in a biospheric context, as he witnesses the inexorable, relentless thrust of life. Fortey writes this story with such a sense of mystery and unknowing that you forgive the audacity of his task and respect the humility he teaches. What is at the core of this humility? And how does humility generate insight?

Searching for fossils has its share of expeditionary glamour, but it is also a painstaking, detailed, and precise form of scientific inquiry. There is much that must be pieced together to resurrect a species, or a community of species, and then to formulate the picture of those species in their environment. You must be aware of a broad literature of paleontological, geological, and paleoecological research. A discovery halfway around the world, in another place and time, may be the key to your work. There is much to decode, uncover, puzzle out, and reconstruct. It is through sheer analytical depth, perseverance, and tenacity that fossil evidence is transformed into a coherent picture of an ancient environment. Even then, as Fortey suggests, you gain only a sliver of knowledge, a narrow view of an inconceivably large picture.

The fossil evidence provides an important scientific foundation for constructing the history of life on earth. But at some point the imagination must also reign. Fortey's vision of ancient life is profusely illustrated by virtue of his imaginary journeys to the ancient environments he studies. His Ordovician daydreaming is a form of time travel, a fantastic immersion into a world that he can never quite know. His imagination fills in the gaps that the incomplete fossil record leaves empty. Imagination and analysis work in tandem—a proliferation of insight occurs when the plunge to Ordovician shores leads to new research ideas. Together they reinforce the unknowing of the quest, yet Fortey is compelled to continue. He is grateful for any glimpse into the past that can be had. "In the last twenty years I have devoted much thought trying to imagine the geography of the Ordovician, piecing together the shadows of vanished worlds. This is like completing a jigsaw puzzle fabricated of scraps and dreams. The fossils have been my guide. I have chased them through jungles and across deserts. I have tracked them along Arctic shores and behind abandoned barns in Wales. Thus fieldwork has kept my feet on the ground and my hands on the rocks. Theories come out of pounding hammer on shale and limestone, conjuring visions of the ancestral faces of continents."[14]

Inspired by an intimacy and familiarity with relic creatures in an ancient era, Fortey's territory moves beyond his immediate location into biospheric realms. What triggers his journeys? It might be the sound of the ocean on the Arctic shore, a stromatolite mat, a low-lying liverwort, or an unexpected fossil discovery on a lonely hillside. These are his tracks to the past, reminders that earth writes its own history. In his most inspired moments, his scientific training fuels his imagination, and he is free to roam and wander.

Beyond imagination and analysis lies compassion, stirred by the capacity to identify with the creatures, environments, and time periods that a person studies. Fortey's Ordovician forays require the fullest possible immersion—to travel in the Ordovician is to become *of* the Ordovician. For a moment he is a trilobite,[15] searching for food in an ancient sea. As Fortey searches the world for fossils, there are moments when the present and past become indistinguishable, and his memory merges with the collective memory of trilobites and graptolites, extinct in body but not in mind, and a window is opened to biospheric perception. This window is also a pathway to praise, reverence, and respect.

Fortey brings a biospheric perspective to his study of natural history. He blends dedicated science inquiry, a lively imagination, and the ability to identify with the critters and time periods that he studies. Hence he is a superb example of an exemplary biospheric naturalist. As an environmental educator, I am interested in the qualities of attention that Fortey brings to his work. Just as he crosses the globe searching the fossil record, finding tracks to deep time, I explore the global change literature for exemplary expressions of biospheric perception. Which global change scientists embody similar qualities in their work? What can they teach us about learning to perceive the biosphere?

In *Symbiotic Planet,* Lynn Margulis challenges readers to consider the origins of life on earth, and to do so by understanding "life from scum." In her "search of the ecological setting of the earliest cells on Earth" she takes her students on a pilgrimage to the remote salt flats of San Quintin Bay in Baja California. They journey to explore the "laminated, brightly striped sediments underlain by gelatinous mud,"[16] what are known as microbial mats. Margulis describes her enchantment with this primordial ooze:

I put my hands in the mud of fragrant microbial tissues and whiff the exchanging gases. Here, as in the human sphere, but neither by commandment nor by necessity, death is part of life. Population growth potential is alternately checked and realized. These seaside communities have persisted for over three billion years. Many inhabitants die daily but the community itself never outgrows its bounds. This is an evolutionary Eden more primal than the greenest grassland. Here, in this earth tissue, animals and plants are all but absent. Even protists and fungi are rare. Mostly bacteria thrive. Standing at the microbial mat, I feel privileged. I delight in escape, thrilled to abandon the urban sprawl of human hyperactivity and exhilarated with the freedom to contemplate life's most remote origins.[17]

Margulis suggests that the origin of life is a mythical concept as it stirs such a deep sense of mystery. "Even scientists need to narrate, to integrate their observations into origins stories."[18]

In her attempt to interpret the evolution of the biosphere, she spends thousands of hours in the laboratory, researching the taxonomic, genetic, and ecological relationships of microorganisms. Through patient and painstaking observation, experimentation, and inquiry, and by virtue of her relentless curiosity and intellectual intensity, Margulis has produced a prolific trail of pathbreaking scientific literature.[19] Yet much of her inspiration is derived from the visceral experience of the microbial environments she studies, her ability to enter them completely, to travel attentively in their domain, at their scale, on their terms. Her laboratory is far wider than test tubes and microscopes. It includes microbial mats, stromatilitic shores, freshwater ponds, and even kefir cultures. She is willing to plunge into the microbial waters wherever she may find them. These unencumbered swims bathe her in reverence and mystery, and are crucial to her perceptual genius.

Tyler Volk, an earth systems scientist, uses a different kind of aperture to the biosphere—hitching a ride on a carbon atom. To interpret earth's physiology, the intricate metabolic pathways of carbon, sulfur, nitrogen, and phosphorus as they move through land, water, and air, Volk provides a travel guide to biogeochemical cycles. This is very helpful, as the movement of an element like carbon from forest to ocean floor and through the soil is not immediately apparent. The biogeochemical cycles entail a scientific understanding of biology, chemistry, and geology— learning the various metabolic equations so as to recognize the combinations and transformations of minerals and life. Yet as Volk demonstrates, a little bit of imagination helps too.

In *Gaia's Body*, Volk portrays the magnificence and complexity of these cycles, describing them as the infrastructure of the biosphere. In the first chapter, "The Breathing of the Biosphere," he explains how scientists trace the amount of carbon in the atmosphere at different times and places throughout daily and annual periods. By studying these patterns, you surmise that the biosphere appears to be breathing. It reveals rhythmic fluctuations, reflecting the respiration of global life. He suggests that the best way to understand "Gaia's web of causes" is to study carbon. At scientific meetings about the carbon cycle, you find plankton ecologists, agronomists, chemical oceanographers, and climatologists, among others, "conversing late into the night" sharing their data about the flow of carbon through earth systems. For Volk, "the humble carbon atom" is teaching scientists how to integrate local research into a new global science.[20]

To make all of this tangible, to make it accessible to people without sophisticated scientific training, requires imagination and identification. Volk suggests following the carbon cycle "to wrap our minds around the entire Earth." So he takes us on such a journey, using metaphor and story. He likens the carbon atom to a mountain lion that rests in a "complex sequence of spaces." He describes the relationship between carbon and oxygen as a "ménage à trois," which "may waft in the air above the seas and continents for years before entering a stomate, or pore, of a grass leaf in the pampas of Argentina."[21]

Volk describes some of the paths a carbon atom travels. It may respire as airborne carbon dioxide, synthesize into chlorophyll, slip into the soil where it's digested as a root-associated bacterium, or travel in groundwater back into the ocean. These routes are just a few pathways in the complex carbon cycle: "In the ocean more stories unfold, the details of which must remain untold for now. But they involve an alga, a tiny swimming crustacean called a copepod, the fall of its fecal pellet, a bacterium a mile under the water, the return to dissolved bicarbonate, then a sinuous lethargic ascent for several hundred years, an equatorial liberation as airborne CO_2, another tour above the seas and continents and then entry into a flower in a Zen garden in Kyoto, which is picked and arranged for a tea ceremony by a practitioner of ikebana and afterwards dutifully laid atop the temple's compost pile."[22]

Volk's strategy in *Gaia's Body* is to teach the biogeochemical cycles by telling the details of their stories. As you follow the carbon atom's journey, you begin to understand the intricate connections between your body and the biosphere. Where does the carbon in your body come from? It comes from everywhere! Volk invites the reader to trace the specifics of its journey and to do so by seeing yourself as the carbon atom, moving through space and time to all corners of the earth, and then finally back again.

I consider Fortey, Margulis, and Volk (along with many others) exemplary biospheric naturalists because their work integrates three pillars of place-based perceptual ecology—analysis, imagination, and compassion. It is instructive to amplify these qualities in more depth, as they are the prerequisites of the *primary methodology* of biospheric perception, the ability to move between conceptual worlds via the juxtaposition of scale. The next three sections explain how these "pillars" are the educational basis for interpreting the biosphere.

Analysis: Using Biospheric Knowledge Systems

To analyze means to dissect, to ascertain the elements of something, to examine minutely. Analysis spurs a series of investigative questions that seek to determine cause and effect. Here, I consider "analysis" in regard to how naturalists learn to systematize their observations. The naturalist looks at a leaf by noticing every detail—noting its shape, color, and form—and then naming its characteristics. The naturalist learns taxonomy to get a sense of structure and classification. This infrastructure serves to organize the collection of data and becomes a point of reference for future observations. Collecting data is achieved through patient observation, record-keeping, and journal entries. Fortey collects and organizes fossils. Margulis develops detailed taxonomic guides to microbial organisms. Volk records the amount of carbon dioxide in oceanic-atmospheric systems.

Sometimes, when analysis becomes an end onto itself it blocks inquiry by providing too much definition. Yet all naturalists require this home territory of knowledge, as a basis from which to set out on further explorations. The birdwatcher must learn how to identify birds according to a common system of knowledge and will eventually commit this system to memory. Naturalists learn systems of knowledge to aid their observations. Identification and systematics are the bulwark of analysis.

What systems of knowledge are prerequisites of biospheric natural history? I suggest three *biospheric knowledge systems*, configurations of information that allow the naturalist to enter the expanded realm of biospheric perception. I offer these systems as substantive contexts for broadening awareness of biospheric space and time, as the basic information that expands place-based environmental learning. Consider these realms as contexts for exploration, guidelines for practice, even conceptual home territories. In each case, the details in these systems proliferate with advances in the environmental sciences, yet they are the ground floor of information for the scientific understanding of global environmental change.

You can't travel through time without a fundamental awareness of the *geological time scale*.[23] This involves understanding the major periods in earth history, the movement of the oceans and continents for each period, and the first appearances of various life forms. There is no other way to appreciate the magnitude of change and the dynamic flows of the biosphere. It is impossible to conceive of evolution and biodiversity without an appreciation for deep time. You can't fully assess the human condition without placing it in a 4.5 billion year perspective. Consider how much public attention is made of the fact that high-school graduates have little awareness of the basic dates of American history, or key regions of world geography. What seems much more egregious to me is the number of "educated people" who have no knowledge of the geological time scale.

The history of life on earth is more readily envisioned by studying the *five kingdoms classification system*—bacteria, protists, animals, plants, and fungi. You cannot fully understand the evolutionary ecology of biodiversity without a full appreciation of the intricate branches of life on the biospheric tree. The purpose of knowing this classification is both to place human life in the context of its ancestry and lineage, and to consider the biospheric function and ecological strategy of each of the five kingdoms.[24]

The relationship between the four elements and biodiversity requires an understanding of the *biogeochemical cycles* of carbon, sulfur, and nitrogen. This involves the ability to follow molecules of these elements through various biospheric spheres (earth, ocean, and atmosphere). Also, it entails tracing the role of the five kingdoms in their cycling. This awareness allows you to understand the evolution of both life forms and their geological environments, and the intricate relationships between so-called biotic and abiotic interfaces. It is a prerequisite of understanding global climate change.[25]

Understanding these biospheric knowledge systems requires diligence, perseverance, and attention. You wouldn't expect to become a good musician without learning the basic structure of your instrument. At some point, even the most improvisational musicians must learn the various scale and chord arrangements of their instrument. Without constant practice, these can't be learned. But just as you finally learn how to play the cycle of fifths without thinking about it, after sufficient practice and study, the geological time scale kicks in too, and awareness of the time periods becomes second nature. Similarly, there's a difference between becoming reasonably proficient at a musical instrument and aspiring to become a professional musician. In each case, you will achieve a different depth of awareness of the knowledge system of the instrument.[26]

The same is true of biospheric natural history. To understand global environmental change requires a base-level proficiency. But without that proficiency, there is no awareness of global change, other than a purely metaphorical and symbolic understanding. Biospheric knowledge systems must be continually referred to and such references should be made in a variety of educational settings with a range of methods (see chapter 8 for an extended discussion of the curricular dimensions of this challenge). This proficiency is merely the first step. It requires illustration and elaboration, and this is best achieved with the other two pillars—imagination and compassion.

Imagination: Visualizing the Biosphere

When you visit a diorama of ancient environments at a place like the Museum of Natural History in New York City, you are merely observing an artist's recreation of that setting as informed by evolution and paleontology, sprinkled with an ideological patina. As Stephen Jay Gould observes in *The Book of Life,* "the social construction of fossil iconography lies best exposed in the conventions that create an enormous departure between scenes as sketched and any conceivable counterpart in nature."[27] But Gould also admits that artistic representations have extraordinary iconic power. Indeed, the very point of *The Book of Life* (subtitled *An Illustrated History of the Evolution of Life on Earth*) is the "fusion of artistic and scientific imaginations to produce encompassing images of the past—scenes from deep time."[28] So the next time you take out your paints to depict dinosaurs roaming among giant ferns, or you feel moved to write a poem about the carbon cycle, keep this in mind.

The imagination is the source of profound insights in both science and art. For imagination is the playing field of aspiration and desire. Sometimes you have to dream a landscape to make it viable. The interpenetration of mind and landscape requires unbounded exploration. To cultivate biospheric perception, you have to dream the biosphere as well as study it. Exemplary biospheric naturalists have the enviable capacity of knowing how to dream and how to study, and how to keep those very different ways of knowing permeable, yet discrete.

So you flip through Gould's book and make frequent stops at the unusual paintings of ancient earth environments. They give you pause, enabling you to reflect and wander, to imagine yourself in the same environment. Here is an important use of the imagination for cultivating biospheric perception—conjure images of what the deep past looked like by drawing them. Lynn Margulis and Dorion Sagan employ a similar strategy in *What is Life?*, portraying the five kingdoms with stunning photography of microscopic worlds. They've gone so far as to develop *The Microcosmos Coloring Book.*[29] Grab your colored pencils and spend a few hours illustrating microorganisms in a desert termite's hindgut, work on the cross section of a diatom, or color some cyanobacteria.

Field naturalists commonly draw birds and flowers as a means of learning how to perceive detail. Why not do the same at different scales? Draw past environments. Color microorganisms. By doing so, you will not only learn the details of their lives and times but you will also let them enter your imagination and dreams. Illustration serves to inspire the imagination. As Gaston Bachelard observes, the imagination "deforms what we perceive; it is, above all, the faculty that frees us from immediate images and changes them . . . Thanks to the imaginary, imagination is essentially open and elusive. It is the human psyche's experience of openness and novelty."[30]

Through the use of the imagination, metaphorical correspondences unfold. In describing marine cycling ratios and the flow of phosphorus from land to sea, Tyler Volk draws an analogy between rivers and blood vessels, describing the universal patterns of fractal branching. He searches for examples of "self-similarity" as a means of traversing scale and contemplating patterns of flow and relationship.[31] "The tiny, riffling arroyos around me would be the river's version of capillaries. As these diminutive tributaries carry substances away from the rivers around them, so capillaries flush wastes from the cells in their neighborhoods."[32] In this way, Volk conjures visions of Gaia's body. He seeks to comprehend and convey biogeochemical cycling through images of biospheric physiology, linking the human body to the body of the planet.

Consider some interconnected uses of the imagination for biospheric perception. The use of *visualization* promotes artwork, poetry, and even music to conjure images of the deep past. By *composing narrative* you might write stories depicting a day in the life of the Ordovician, or the adventures of a carbon atom. You *juxtapose scale* to enter microscopic domains or to track the movement of continents. *Creating metaphors* conjures correspondences between biospheric knowledge systems and everyday life observations (Gaia's body, microbes as planetary elders).

Compassion: Identification and Reverence

Even the most imaginative insights and the most comprehensive scientific study can only yield glimpses of biospheric awareness. The depth and complexity of these connections are way beyond the perceptual and intellectual capacity of any individual. What exemplary biospheric naturalists also hold in common are feelings of humility, praise, respect, and reverence for the grandeur of the biosphere. These qualities are the foundations of compassion—an ethic of care for the fabric of biodiversity, for the whole Earth project, beyond the chauvinistic needs of the human species. This sensibility is sustained through identification and empathy, the ability to see oneself embedded in the biosphere. Consider Tyler Volk's concluding remarks in *Gaia's Body*.

I will sign off soon, giving my mind a much needed break, while I step outside to a glorious sensuous awareness of the connections among the vast atmosphere, the new spring leaves, the depths of evolutionary time, and my own humble breath. Inhale. Exhale. Through this breath I am connected to the deep ocean, the burial of detritus, the cycles kept spinning from year to year. I am connected to the yellow lichen on the tan bluff and floating foramaniferans on the other side of the world. I am connected to Thioploca, a corn plant, a nematode; to nitrogen-fixing nodules, the tundra, the stratosphere. Gratitude is heaven, and heaven is surely here on Earth. Thus I say, thank you, Gaia.[33]

Or the final remarks in *What is Life?* by Lynn Margulis and Dorion Sagan:

Our destiny is joined to that of other species. When our lives touch those of different kingdoms—flowering and fruiting plants, recycling and sometimes hallucinogenic fungi, livestock and pet animals, healthful and weather-changing microbes—we most feel what it means to be alive. Survival seems to require more networking, more interactions with members of other species, which integrates us further into global physiology.[34]

And, finally, in *Life*, Richard Fortey, after sketching the bare outlines of the natural history of humanity, ponders how a "review of the history of life should provoke awe, above all else."[35]

Imagine if a narrative as complex were known for all the other millions of species living in this tangled and prolific biosphere; why, the paper used to print their histories would strip the trees from the world! Yet the chances are that every species has a story worth telling. In your mind's eye imagine you are an eagle gliding above the rain forest canopy, and as far as you can see in every direction, there are billowing canopies of trees reaching upwards into the light. Each one of these trees might represent the branching history of a living species, and the forest itself might be a crude representation of the density of the historical past. We shall never know every detail of every tree, but we can understand the vitality of the whole, and thus see the forest for the trees.[36]

These passages taken together reveal the unfolding of a master narrative of the biosphere. These naturalists aspire to experience what I think is at the core of biospheric perception—the unfolding of one's life in relationship to the history of life on earth. To observe biogeochemical cycles, to trace the lineage of biodiversity is to observe biospheric creation. The story of your time is the story of all time. Hence the three pillars—analysis, imagination, and compassion—work in tandem as approaches to experience and awareness.

Recall the primary emphasis of this chapter—how can biospheric perception become intrinsic to everyday awareness? There are four qualities of attention that are helpful in this regard—patterns of awareness to track environmental change. First is to attach *mindfulness* to the biospheric dimensions of everyday activity. One can learn to breathe with the biosphere in mind, or to practice specific meditations for doing so. Even an act as banal as recycling can be a reminder to consider biogeochemical cycling. Turning on the faucet to get a drink of water can remind you of the hydrological cycle. No one can think in these terms all of the time, but it is still a useful perspective to encourage. Second is the faculty of *serendipity*. Observing the biosphere may entail being in the right place at the right time. There are special moments—unusual weather phenomena, encounters with wildlife, celestial occurrences, outstanding views—which call attention to biospheric processes Third is to be aware of the *immanence* of the biosphere. Biogeochemical cycles occur whether you pay attention to them or not. You don't need to go anywhere or do anything special to acknowledge the ubiquity of these processes. Fourth is to *observe patterns*. This includes *noting difference,* observing discontinuities, making special note of patterns in nature that seem different from the usual and then exploring the meaning of that difference. By *noting similarity* you observe patterns that occur with regularity. Difference and similarity are functions of familiarity. The more you observe a landscape, the easier it will be to spot the expected and unexpected.

Here is a challenge. From your backyard, can you explore your surroundings, field guides in hand, with charts depicting the geological time scale, the biogeochemical cycles, and five kingdom taxonomy easily available, ready at the call to move through deep time and travel through the biosphere? What additional preparation do you need? The next sections of the chapter provide specific examples of how to do this. We'll consider how to interpret the biosphere by observing the patterns of environmental change—tracking thresholds, weather, cycles, and lineage and ancestry.

Tracking Thresholds

Several Junes ago southwest New Hampshire was drenched in a month of continuous tropical rain. The early morning sun inevitably yielded to booming midday storms which continued throughout the afternoon and evening. These were powerful and quick rains with intimidating thunder and lightning.

I live on a dirt road in the forest. Although the road receives its fair share of wear and tear—the potholes are outrageous during mud season—the road has been passable for the twenty-five years I have lived here. You assume that a road is always there, ready to hold your vehicle when it's time for you to go someplace. One morning, during this June stretch of rain, after an evening of particularly heavy thunderstorms, I was about to leave the house when I noticed that the road was flooded. The accumulated sediment from the nightly rains had overwhelmed a nearby culvert, diverting a local stream into the road. Coupled with the runoff from the rain and the already saturated forest soil, the stage was set for havoc. What was once a road was now a river. Like a kid at a gushing fire hydrant, I was thrilled at this scene. I took off my shoes and socks to wade in the knee-deep water, entering the flow of the flood, following it to its source. Several neighbors joined me and we deliberated on the cause of the flood and how the road could be fixed.

This small flood was an instant lesson in fluvial geomorphology. You could observe the formation of small-scale canyons and valleys, the creation of meanders and examples of alluvial deposition. At an accessible, directly perceivable scale you could see how water transforms landscapes. A miniwatershed was forming before my eyes.

After working for most of the morning the road crew repaired the culvert. The raging brook was reduced to a thin trickle. Eventually the road was dry. It was massaged by big dirt-moving machines and by the next day the road was back to normal. The casual observer or someone passing through for the first time would never have known what transpired. Yet today as I inspect the scene, I can detect evidence of the miniflood. At the end of our driveway, just in front of the stone wall that separates the road from the woods, there is an outwash deposit of sand. Most of it came on the day of the flood. It's only ten feet long by three feet wide, but it is nevertheless a reminder that for twenty-four hours a brook flowed here, depositing a layer of silt and sediment on the forest floor.

How can this miniflood be placed in a biospheric perspective? First, it's useful to think about the spatial and temporal scale of any event. The weather pattern that shunted tropical lows to New Hampshire, contributing to these storms, lasted for nearly a month, and covered several thousand miles of Atlantic coastline. The storm front which caused the miniflood stretched for several hundred miles and lasted for four hours. The thirty-square-foot sand deposit may stay here for a few hundred years, unless some other geomorphological process, which might include human interference, causes additional changes. So a short-term weather system, of extremely limited duration, caused a small, but noticeable change in the landscape that may be evident for several human generations. An atmospheric event caused topospheric (landscape) change.[37]

Playing with scale in this way makes it possible to interpret longer-term processes. It's much easier to understand how a significant flood can alter the landscape for centuries. Recall the incredible Mississippi river flooding of several years ago and how it turned much of the American Midwest into an inland sea. A flood, regardless of scale, is a biospheric process, linked to the movement of water and energy across the earth. Remember too the impact of Hurricane Mitch in Central America. It caused massive human and ecological damage.

Environmental change may be traced to the complex chain of events triggered by an extraordinary event. Or it might be the accumulation of thousands of barely perceptible minievents which accumulate over time. Many environmental changes are a combination of both. They are interconnected, multiscaled processes. That's why it can be so hard to attribute causation to any particular trend. Nevertheless, it's illuminating to consider the patterns intrinsic to multiple scales of environmental change. Finding such patterns is a good way to consider the scale of biospheric processes.

For example, why did the road in front of my house turn into a river? A *threshold* was reached which allowed a stream to overrun its banks. The result was a small-scale geomorphological process that I could witness over a very short period of time. Interestingly, I didn't actually observe the moment when the limit was breached. Thresholds are crossed as a consequence of seminal episodic events that have long-lasting, even catastrophic effects. Often, they are not directly observed.

Geographer Richard Huggett, in his fine text *Environmental Change,* describes the crossing of a threshold in broader temporal terms. He suggests that were it not for a cooling and aridification trend during Tertiary time, modern humans, in the form we know ourselves, might not exist. Around 2.5 million years ago, the modern ice age was triggered when atmospheric cooling reached a threshold and the Antarctic ice expanded. With the onset of the ice age, in response to drier conditions, savannas replaced woodlands throughout Africa. Grassland species replaced forest-dwelling animals. These conditions allowed for the emergence of the human genus who became adapted to grassland habitats. "Directional changes (trends) often occur when a threshold within the ecosphere is crossed . . . Paradoxically, the sapient species whose origin was occasioned by a global environmental change is now changing the environment on a global scale."[38]

It's instructive to search for examples of thresholds in your place and to use them to extrapolate biospheric processes. At what point does local development directly impinge on the habitat of an endangered species? How does the removal or planting of trees change the availability of carbon in the proximate environment? These are crucial questions for biodiversity studies. You may not be able to witness the exact moment when a species is threatened, when the "threshold" is crossed, or what limiting factors are most crucial, but knowing there are measurable thresholds provides you with useful clues in assessing a habitat.

Considering thresholds is a way of thinking about environmental change. The same thought processes that you can employ at a local scale can be used to consider these broader questions. Granted, sophisticated instrumentation is often required to detect the evidence of environmental change, but that is not always the case. If you know what you are looking for, there is much that can be detected with the naked eye, on a Thoreauvian scale. Observing the patterns of environmental change, such as the crossing of a threshold, helps you understand the relationship between local and global processes, and the more you detect these patterns, the more feasible it is to make broader conceptual connections.

Tracking the Weather

You can detect the tracks of environmental change using the weather as a guide. If you gaze at the sky and watch a cumulus cloud form and then disappear, or perhaps build to an ominous thunderhead, you are tracking the flow of moisture through the sky. If the cloud builds sufficiently to form rain, you can track the movement of moisture from heaven to earth. You can feel the cool rain as it falls on your body. You can watch it as it hits the earth, flowing through the landscape, carving channels, carrying detritus, forming runnels and pools. Later, when the sun comes out, if the day is warm enough, you can watch the fire of the sun evaporate the pools as the moisture ascends back to the atmosphere. Here is the hydrological cycle, the dynamic interplay of water, air, fire, and earth moving through your local place.

To move beyond your place, you can track the storm backward and forward in time, as it traverses through space in the biosphere. What is the cause of today's weather? Perhaps a cold front has come barreling through your region, carrying cold, Arctic air, generated over the poles and now careening toward the humid subtropical southeast. Or perhaps thick, humid, saturated tropical air is spewing northward out of the Gulf of Mexico, bringing rain to the cold, northern boreal forest in the middle of winter, melting the snow, hastening the arrival of spring. Depending on the circulation of the atmosphere, the Gulf of Mexico may send waves of moisture to the American Southwest causing the summer monsoon, as thunderstorms break out over the Colorado plateau.

If you look at today's weather map, you can watch the highs and lows, the warm fronts and cold fronts, the tropical and Arctic air masses, and see how maritime Pacific air will travel across the country and eventually drop rain in Boston. Several days later, that same air mass, modified by the Gulf Stream, will bring showers to the British Isles. The air you are currently breathing has taken a long journey across oceans and continents, over mountains and great lakes, through cities and canyons, until finally it passes through your body. Tracking the weather is a fine way to observe the four elements as they embody air and water circulation and the movement of heat energy as it envelops the landscape. What air are you breathing right now? Where has it been for the last week? Where will it travel tomorrow? Follow the weather as it moves around the world. By doing so, it's easier to envision various global circulatory systems.

The seasoned observer can "read clouds" just as the ecologist can read the landscape. A cloud formation tells a story of air motion, moisture, and topography. On one scale, the cloud helps you track a weather system as it moves over the landscape. At another scale a cloud represents a microscopic world of the four elements transforming—water turning to ice, particles forming, liquids, solids, and gases embroiled together. On a third scale, observing clouds is a clue to following oceanic and atmospheric circulation—primary conveyors for the biogeochemical cycling of carbon, nitrogen, phosphorus, and sulfur. The deeper you look into a cloud, the more substance you may see in its ephemerality. The cloud is a tracking device at several scalar levels.

As you become more familiar with cloud formations, you realize that specific formations are indigenous to particular weather phenomena. There are subtleties and nuances even in the way, let's say, cumulus clouds will form. Similarly, there are indigenous weather patterns that one can come to expect over time. Every place has an annual rainfall—the amount of moisture that the planet dumps in an area over a year, an annual cloud cover, and, of course, average temperature ranges. You can explore the transformation of the four elements by noting how much heat, moisture, and wind moves through any landscape.

Observing the weather has always been crucial for people who earn a living on the land. You can't always predict the weather—there's too much uncertainty and way too many variables—but it's useful to recognize patterns of drought and moisture, and to interpret the ways that the weather changes in relationship to other processes of natural history. If you know what to expect, you have a much better idea of how to grow or find food, where and when to seek shelter, whether it will be a good or bad year for blueberries.

Learning how to observe clouds and weather patterns is a fine way to interpret the biosphere. When you experience an unusually wet and mild winter you can ask some questions about the prevailing patterns. Why is there so much additional heat and moisture here this winter? Where do you look to find the origins of this situation? First, you consider the broader oceanic and atmospheric circulation systems. Where is the weather generated and where is it coming from? This requires having access to climatological data. Satellite maps that indicate the temperature of the oceans reveal that this winter's weather is directly attributed to the El Niño effect, the cyclical warming and cooling of the Pacific Ocean. By observing such climate cycles, you can see how this year's weather fits into a broader circulatory pattern. Second, you try to understand the regularity of pattern. What phase of the cycle are you observing? Is the El Niño effect more or less frequent? Is its manifestation more or less extreme? For what time scale do you make this assessment? Third, you consider the causes of the pattern. Why does the temperature of the Pacific Ocean fluctuate from one year to the next? How does its fluctuation correspond to other atmospheric cycles?

Cycles of climatological causation are biospheric in scope, whether you are observing short- or long-term patterns. Today's weather is not only a result of the prevailing circulation of air masses but also a response to a series of long-range patterns—tectonic activity and the movement of continental plates, variations in the Earth's orbit around the sun, balances in the biogeochemical cycles. Increased levels of carbon dioxide in the atmosphere cause the "greenhouse effect." Heat is trapped within the trophosphere, contributing to a global warming trend. Processes such as the burning of fossil fuels and widespread deforestation release carbon. When you drive to work today, or read the newspaper, your actions indirectly contribute to tomorrow's weather. You can't trace the sequence of events because the impact of your actions is seemingly trivial, but you can follow the conceptual chain and link your everyday behavior to an expanded cycle of causation.

Just by thinking about this conceptual chain, the idea of global warming becomes more tangible. You start by noticing difference (unusual and persistent changes in heat or precipitation, a change in amount or severity of storms, a switch in prevailing wind direction), and trying to locate the broader patterns that help explain the difference. As you move through multiple scales, you realize that climatic changes result from an assortment of factors, and anthropogenic influences are just one of many causes. As you study the past you recognize the dynamic changes in local climate and the time it takes for these fluctuations to occur. Rates of fluctuation have a context. A three-degree change in annual temperature over the course of a century may have a threshold effect. It may catalyze a series of unprecedented changes in a short period of time—sea level rise, extreme weather, or unusual droughts. How might previous glaciations and interglacial warming trends inform us about such fluctuations?

This temporal context places your anecdotal experience in perspective. How do you know where today's observations fit into a broader cyclical process? Several hot summers don't prove global warming. But a century-long increase in temperature or a growing number of severe hurricanes might. The patterns of local change may reflect longer-term trends, but you can't know this without comparing data over a variety of scalar levels. To interpret the biosphere, you start by closely observing what is near at hand, and then you compare your observations to broader patterns of environmental change. Today's weather is a response not only to yesterday's circulation but to the dynamic evolution of the biosphere. To contemplate global warming, you interpret the biosphere by tracking the weather across space and through time.

Following a cloud around the world, tracing the origins of the air you breathe, tracking today's rain to a distant ocean, noticing subtle microclimatic variations—these are ways of using place-based perceptual ecology to interpret the biosphere. You come to know your place by observing the details of the local weather, and as you do so, you expand your range to include the entire globe. Your place is no longer just an enclosed landscape. It is also inextricably connected to a complex of global circulatory systems. By following these cycles of circulation, you learn how to track the patterns of environmental change.

Tracking Cycles

As I sit on my porch on this humid, summer morning, I am participating in dozens of biospheric cycles, although I am aware of only a few. The changing of the seasons is easy to observe. The weather is hot and muggy today, and the forest canopy is rich and dense, but I know from direct experience that in only two months, the leaves will be changing color, the days will be much shorter, and the weather will be crisp and cool. It's harder to follow the biogeochemical cycles because they occur on such diverse scales. Yes, as I breathe in and out, I am aware that I am recycling carbon atoms, that the molecules that pass through my body are intrinsic to a complicated flow of energy and matter throughout the biosphere. But it does take a special effort to remind myself of this. I also know that the interglacial warming that allows for such a warm, humid summer at this latitude is a consequence of the Milankovitch cycle, an

astronomical relationship between the earth's wobbly orbit and its proximity to the sun. By paying attention to the night sky, and observing
ionospheric activity (Northern Lights), I know that there are sunspot
cycles.

As a perceptual experiment, it's good practice to try to identify cycles
in your proximate environment, to identify the phase of a cycle that you
are observing, and to try to conceptualize aspects of the cycle that are
both visible and invisible. Observing cycles is a fine way to track environmental change. It's a good way to assess trends and discontinuities.
Is the phenomenon you are observing an expected sequence in a regular
process or is it an anomaly? How do you know? Consider keeping a *journal of cycles,* with the express intention of using your local observations
as benchmarks for considering the phases of the cycle that are beyond
your space and time, using the instructional aids of biospheric inquiry to
expand your gaze.

Ecologists Allen and Hoekstra suggest that "if one waits long enough
almost all processes that at first appear to be a linear progression will
emerge as cyclical."[39] They present an example from fire ecology. Although a forest fire might seem like a "one-way process of destruction,"
it serves an ecological function by opening the vegetation so that plants
which require high light levels can return, thus rebuilding fuel levels for
the next fire. This raises an interesting question for ecology researchers.
Any research project, depending on its time and space limitations, may
only represent a small phase of a much larger cycle. It's useful to keep
this in mind as you try to track environmental change.

Allen and Hoekstra propose that the idea of a cycle serve as a unifying
concept for ecology, a means of finding pattern in complexity. They suggest that populations, communities, ecosystems, and landscapes represent the "embodiment" of cycles. Likening these cycles to a game, they
observe that "players in more than one ecological structure occur in
more than one ecological cycle." Players may be involved in several
games simultaneously, places where diverse cyclical phenomena inter-
sect. "Those structures are the places where the various cycles of nature
kiss."[40]

This concept reminds us of the utter complexity of interpreting the
biosphere. Any place-based observation, if pursued diligently, leads
down a trail of inquiry that may be way beyond the grasp of the lay observer. Yet you have to start somewhere. You can't begin to perceive
global environmental change unless you pay attention to the intricate
details of the place where you live, and then begin to ask some broader

questions so that your daily observations are connected to biospheric dynamics that aren't immediately obvious. Hence the importance of pattern recognition as a perceptual tool. When you learn how to track environmental change in your own backyard, you can adapt the same observational processes to broader scales. The challenge for all observers is to interpret the correspondences between scales by identifying patterns.

Tracking Ancestry and Lineage

Another way to track environmental change is to reconsider your family history. When was the last time you looked at your birth certificate? It's a straightforward document with the time and place of your birth—a bureaucratic proof of existence. Most people enjoy birthdays, celebrating them with showers of gifts, making special note of dates when their family and closest friends were born. Nothing is more embarrassing than forgetting someone's birthday!

What if birthdays were occasions to interpret the biosphere, and to do so from the perspective of your evolutionary origins? On your next birthday, create a *biospheric birth certificate,* tracing your lineage through geological time. Most people can trace their ancestry back a few generations, at least long enough to establish their ethnicity, the country they come from, and perhaps a few other details. Given the 4.5-billion-year history of the earth, and the 3.5-billion-year legacy of life, it's astounding how little we know about our lineage. Once you've recalled a few folks from the "old country" it's hard to dig much deeper, let alone recall the ancestry of *Homo sapiens,* the origin of mammals, or the arrival of life on land from the sea.

If you create a giant mural to serve as a biospheric birth certificate, how might you pay homage to the full spectrum of your lineage and ancestry? Where would you even begin? It's always useful to search for the past in the present. You can start by observing the living representatives of the five kingdoms, gaining the full scope and splendor of contemporary biodiversity.

Lynn Margulis and a team of writers provide three excellent complementary resources for this search. Start with the superb field guide *Diversity of Life.* The book is organized in five chapters, each corresponding to a taxonomic kingdom. Each chapter suggests locations where you can learn how to observe members of the kingdom. For bacteria, there are sections for garden soil, pond scum, woodland streams, foods, and rocky brooks. For protists, you can look in swamps, fallen logs, outflow pipes, or decaying pond vegetation. Look for fungi in forest clearings or orchards.

Many of the species covered are microscopic and you need special equipment to see them. Still, it's instructive to visit the habitat, field guide in hand, to observe where the microbial organisms live. With some practice, you can find examples virtually anywhere of species from each of the five kingdoms. Of course, locating plants, animals, and fungi is easiest to do, whereas learning which microscopic species inhabit a place takes research and practice. I require the *Diversity of Life* field guide when I closely observe the green soybeans that are just about ready for harvest. I read about the phylum Proteobacteria, the nitrogen-fixing aerobic bacteria. "The best-known nitrogen fixer is *Rhizobium,* which forms a nitrogen-fixing symbiosis with the root hairs of plants from the pea family (Leguminosae), such as soybeans, alfalfa, clover, and lentils, These bacteria provide one of the crucial elements for plant growth, nitrogen, by fixing the inert gaseous nitrogen (N_2) of the air and transferring it to the carbon compounds of the bodies of organisms . . . Without the oxygen-sensing nitrogen-fixing capability of these bacteria, which must have an ancient and venerable history, we would all starve from protein deficiency because of the lack of available protein-forming nitrogen."[41]

The rhizobia transform in the process of nitrogen fixing, forming a symbiosis with the soybean plant. They form nodules on the root of the plant as the symbiosis matures. Later in the fall, when I harvest the soybeans and pull the roots up from the soil, I can easily observe these nodules. The complicated and elegant biogeochemistry of *Rhizobium* is beyond the scope of this chapter. Rather, what I wish to convey is how this place-based observation is a way to trace ancestry and lineage, and in the process a means of interpreting the biosphere.

First, with the help of field guides and other conceptual aids (microscopes) you can observe the diversity of life in your backyard. Try to observe as many representatives of the five kingdoms as you can. If there is a fallen log near the garden, you're likely to find protists. Surely you can find animals, plants, and fungi. In a place as common as a garden you can observe a variety of phyla. The *Diversity of Life* field guide makes you aware of life's microbial layers, so near at hand, but so subtle to discern—the invisible vibrations of life referred to by Vernadsky. It is at the microbial level that the intricate connections between organisms is most clear, especially the symbiotic relationships that are crucial to ecology and evolution.

Second, find the closest relatives you can to the species that you are observing, and consider the varieties of niches they fill. When I look up Proteobacteria in Margulis and Schwartz's *Five Kingdoms,* I note that *Rhizobium* is one of 104 genera! "Among the many different species of phototrophs, some are tolerant of extremely high or extremely low temperatures or salinities. In each group, some kinds are capable of fixing atmospheric nitrogen."[42] The soybean is in the Anthophyta phylum, subphyla Dicotyledones, the largest group of flowering plants, comprising 170,000 species.

Third, use *Five Kingdoms* to look up the evolutionary path of the species and find out when it emerged on the geological time scale. The first bacteria probably appeared at least 3.4 billion years ago. The Proteobacteria (or purple bacteria) are among the oldest of bacteria, about as old as any living creatures. The evolution of *Rhizobium* is harder to trace, and its symbiosis with the Leguminosae family (beans and peas) is presumably a much more recent phenomenon. The first land plants date to the Silurian period (410 million years ago), and the flowering plants are among the youngest of plants (120 million years ago). The soybean is a domesticated plant, first cultivated in China about five thousand years ago. By contrast, the first mammals appear in the Triassic period (210 million years ago) and the appearance of the first hominid ancestors is about 20 million years ago.

Fourth, consider the role that microbial species have in contributing to the functioning of the biosphere. By virtue of "biofixation," *Rhizobium* plays a vital role in the nitrogen cycle. Nitrogen is an important limiting factor in the productivity of terrestrial ecosystems. In *Cycles of Life,* Vaclav Smil describes the role of the biogeochemical cycles in biospheric systems. He remarks how life depends on the ability of a surprisingly small group of living organisms to fix N_2 into a usable form.[43] Rhizobia make the most prominent contribution on a biospheric scale, although 60 genera of cyanobacteria are the primary nitrogen fixers in the oceans and in the Arctic. See Smil, Margulis, and Volk for more detailed discussions of the role bacteria play in biospheric processes.

In *What is Life?* Lynn Margulis and Dorion Sagan write lengthy essays which describe the role each of the five kingdoms plays in evolutionary ecology and biospheric natural history. For each kingdom, they pose the fundamental question, "What is life?" In regard to plants, "life is the transmutation of sunlight . . . It is the energy and matter of the sun become the green fire of photosynthesizing beings. . . . Green fire converts wildly to the red and orange and yellow and purple sexual fire of flowering plants."[44]

It's interesting to develop a five kingdoms species inventory of the place where you live. Quickly you realize the impossibility of performing such a task comprehensively. It's hard enough to learn all of the visible plants and animals, let alone to name and observe the soil microorganisms, the bacteria that live in mammalian guts, the microscopic symbiotic species, or the spores that float through the air. But you can learn about biodiversity by appreciating the full scope of ecological and evolutionary possibility. When a wetland is filled, it's not just the more visible and "popular" flora and fauna that are threatened. Equally significant is the enormous catalog of microbial species that evolve with the habitat. You won't miss them until you know that they are there.

As Margulis, Schwartz, and Dolan suggest, "we are clearing out our planetmates faster than we are learning about them."[45] A five kingdoms perspective broadens awareness of this threat. Each planetmate is a legacy of rich natural history, in many cases hundreds of millions of years old. Even within your human body, you carry the legacy of several billion years of biospheric evolution. Margulis describes cells as biochemical museums and microbes as planetary elders. There are clues to the history of ancient earth in your body.

The evolutionary paths and ecological trails that lead to your local place are far more complex than you can ever know. A five kingdoms perspective contributes to an *appreciation* for biodiversity. It enables you to trace the history of life on earth that leads to this moment. To become more familiar with the other species in your habitat, it's essential to interpret those species from a biospheric perspective. That allows you to glimpse their legacy, the remarkable story of how they've come to share this place with you, the role they play in maintaining its ecological integrity. In almost every case, you (human) are the newcomer. The stories of these species, whether it's through tracing their ancestry, or understanding their role in biospheric functioning, provides a full picture of what it means to live in a place. Perhaps this is the special responsibility of humanity by virtue of our symbolic capacity—to serve as planetary record-keepers, to witness biospheric creation as it occurs at this moment.[46]

Barefoot Global Change Science

To interpret the biosphere, I propose a barefoot global change science. Cadres of citizens, schoolchildren, elders—people from all walks of life—meet in schools, libraries, parks, and on the Internet, to share stories and data. They pool their observations and expertise so they can track environmental change in their neighborhood. Via electronic communications, they compare data with folks from other places. Professional environmental scientists work regularly with citizen groups and schoolchildren to provide training and guidance. They jointly establish local research projects. Artists draw biospheric murals on the sides of buildings. A special television channel shows global change satellite maps twenty-four hours a day. Every computer is sold with built-in geographic information system software.

Perhaps this is a naive dream. Yet I am sure that grassroots, hands-on, participatory, place-based learning is the best way to learn about global environmental change. This approach is "barefoot" in several ways. First, it can be accomplished in your local place for however long you may live there. The garden soil, an outflow pipe, a coastline, a city park—anyplace where there is life—is a place to observe the biosphere. Second, it must be achieved collectively, through a concerted community effort that builds support, encouragement, and pools expertise. Third, it is multigenerational and multicultural. People of different ages and different perspectives have special perceptual skills that reinforce community learning. You can't interpret the biosphere by yourself. The notion directly contradicts the very processes you observe. This barefoot effort is generated in thousands of localities, but it must be truly global in scope. It is the entire biosphere that we are studying. It doesn't privilege any one place or species.

This collective project requires enormous amounts of data, much of which is collected locally, through place-based environmental learning. But it also requires sophisticated instrumentation—technologies that stretch perception through magnification and miniaturization. It demands fine maps, some made by hand, some photographed from the sky. It is catalyzed by efficient communication—the movement of vital data through electronic means. What is the best way to balance place-based perceptual ecology with advanced electronic communications? These technologies have a dual potential. In some ways they are powerful perceptual aids helping you see dimensions of the biosphere that you could barely even imagine. Yet they also serve to inhibit perception, taking your senses away from the land in favor of the screen. Cultivating biospheric perception requires urgency and balance, speed and deliberation—the Internet, electronic maps, and binoculars, coupled with the naturalist's gaze. The next chapter considers how such a balance might be achieved.

6

The Internet, the Interstate, and the Biosphere

Information, Speed, and Distance

Electronic communications and high-speed transportation both enhance and diminish biospheric perception. By virtue of your participation with various technologies of information, speed and distance,
you are engaged in a remarkable perceptual experiment. Once you've
watched television, surfed the Internet, driven on the interstate, or taken
a plane trip across the country, you never again see the world in the same
way. Yet such experiences quickly become routine. How easy it is to forget the stunning impact they have. The Internet and the interstate are
powerful networks whose speed and ubiquity dramatically transform
both pace and perception. In what ways do these networks promote and
inhibit awareness of global environmental change?

My premise in this chapter is that technologies of speed, distance, and
information, if used judiciously, have great promise as learning tools for
biospheric perception. But their use must be balanced with the delibera-
tive gaze of the naturalist. Is it possible to move seamlessly from the Internet to the forest and back again? What risks and opportunities are
unveiled? I suggest that a place-based pace of observation is the foundation from which faster speeds can be more safely explored.

Marshall McLuhan was responsible for one of the great insights of the
twentieth century when he declared that the "medium is the message." I
have always taken that to mean that the perceptual process of watching
television, for example, is much more significant than whatever program you are watching. Television teaches millions of people how to
scan and observe images and pictures at rapid frame rates. Over five
decades of television has exposed viewers to increasingly faster bytes of
sound and sight.

The remote teaches people how to move from program to program, preparing them for the point-and-click interface of hypertext. Many television viewers watch multiple programs simultaneously. You learn to scan the airwaves. This skill prepares you well for the Internet where you hop from home page to home page, flipping through images and text at great speed. In a twenty-minute session on the Internet, if my server and computer are fast enough, I can easily visit several dozen home pages, glance at assorted e-mail messages, thereby surveying my information territory, noting areas of interest that I can visit in greater detail if I so choose. I've learned how to move pretty darn quickly, and to do so rather comfortably. This scanning skill, a modern form of hunting and gathering (but for information instead of food), is a matter of survival for Internet users. Purposefully scan or aimlessly perish!

There is a similarity between observing the fast-moving images of the Internet and watching the world fly by on the interstate. When you travel in a car or train, you get accustomed to moving through lots of scenes very quickly. In just minutes, you may observe residences, shopping districts, and parks, as assorted communities, buildings, and habitats pass by. This is even more pronounced in an airplane, when the speed of transport covers great distances and in one fell swoop you are whisked to another part of the country. If you pay any attention to what you are flying over, you notice remarkable patterns of settlement and landscape. On a transcontinental or overseas trip, you move through entire weather systems and ecosystems in several hours.

Both the Internet and the interstate, broadly conceived as structures of information and transportation, are networks of perceptual change. The Internet conveys information. The interstate transports people and goods. Their efficiency relies on technologies of speed.[1] The Internet and interstate, traveled together, provide "access" to many places at once. Speed is the means of access. What is most striking about one's use of these networks is how speed transforms scale. Speed alters the perception of space and time. Fast and slow, big and small, near and far—these are relative qualities, dependent on speed. Everyday use of the Internet and the interstate is a perceptual experiment in the alteration of space and time, through the shifting of scale.

In chapters 4 and 5, I showed how by paying careful attention to the natural history of your home place, you could slowly expand your vista. With various perceptual aids, you could juxtapose scale, and thus explore further in space and time—learning how to perceive biogeochemical processes, the movement of weather systems, and the ecological and evolutionary context of your home place.

What's remarkable about technologies of information, speed, and distance is that they accelerate this process, enabling you to scan and manipulate broad realms of data. With computers and the Internet you have access to visual portraits of complex systems, comprehensive mapping, international data collection, and the global exchange of local information. Watch birds migrate, fires sweep the Amazon, and storms cross the Atlantic from your laptop! With high-speed transportation you can study diverse habitats in far-flung corners of the world, attend international conferences, or view the world from thirty thousand feet. You can use the Internet and interstate to cover broad regions of conceptual territory, comparing biomes and ecosystems, doing so from computer screens and airplane windows. What a fine opportunity for conveying images and ideas about the biosphere.

Yet there are also seductions and traps. Every hour spent in front of the computer screen comes at the expense of your time observing natural history directly. There is a profound difference between the rapid pace of fast-moving pixels, the anticipation of gathering information as fast as your server allows, versus the deliberate gaze of natural history observation. Too much time spent in cars, trains, and planes alters your view of a landscape so that you only see what is passing you by. You may never slow down enough to see the details in the portrait.

When you pass quickly through multiple scales, you have the opportunity to interpret and detect patterns that you may not see at a more deliberate pace. Here lies both potential and jeopardy. For example, in traveling so quickly through so many places, is your vision of home place expanded or obliterated? Are you so absorbed in the pace of travel that you learn to savor speed simply for its own sake? Or do you use speed as a means of comparing perspectives, gaining further appreciation for the details of what lies close at hand?

The Internet and the interstate strike me as the psychedelics of biospheric perception. They provide remarkable vistas at an extraordinarily rapid pace. They open the doors of perception so that you can take in much more than you may be prepared to see. Their perceptual impact is veiled by their habitual use. The world moves forward by leaps and bounds and you are always trying to catch up.

The first step in understanding the impact of these technologies is to explore the perceptual implications of their use. The most direct and effective way to achieve this is by reflecting on your own experience with technologies such as televisions, computers, and cars. How have they changed your view of the world? I stress this approach because by comparing your experiences of the world, with or without, let's say computers, you gain some terrific insights into how they alter perception. The first section of this chapter is a technology memoir, an educational approach I advocate as a means to reflect on one's experience of technology. This approach serves both to remind you how different the world looks in the presence and absence of a particular technology, and also to understand the extent to which you embrace or abhor its use, or what's most likely, how you might deal with your ambivalence.

The second section, influenced by the work of ecologists Dan L. Perlman and Glenn Adelson, suggests the virtues of place-based paces as a cognitive foundation for understanding speed and scale. Depending on your pace and its inherent scalar perspective, different qualities of biodiversity are revealed or concealed. The third section surveys some of the interesting perceptual opportunities provided by computer mapping, global information exchange, and the Internet, including interesting websites that enhance one's understanding of the biosphere. What unique perspective do they provide? The final section advocates specific perceptual checks and balances, emphasizing the importance of using caution and judgment while experimenting with speed and scale.

My view is that one's perceptual relationship with technology is pertinent to understanding biospheric perception. This relationship matters because most people spend so much time watching television, using computers, and traveling frequently in cars and airplanes. Despite the best efforts of environmental educators, millions of people mainly observe nature through windows and screens. Nature is something you walk through to get from one building or vehicle to another. The modern commuter moves from a swiftly moving vehicle to the Internet and back again, perhaps sandwiching enough time for a brisk walk. And even if you make an effort to spend time outdoors, or you work in an outdoor profession, unless you completely eschew electricity, you spend much of your time working with, being entertained by, and transporting yourself in machines.

A place-based perceptual ecology is only relevant as it incorporates this reality, for it's through machines that most people observe the natural world. That's why it's so important to reflect on the impact of the machines we most rely on as perceptual aids. It's much different viewing the biosphere through a computer than it is from the top of a mountain. Yet the permeability of the visceral and virtual is the hallmark of our age, so it is worthwhile to reflect on their integration *and* separation, and to take perceptual advantage of both milieus.

Technology Memoirs

I remember sitting in my room in the corner of our Queens, New York apartment house, engaged in whatever four-year-olds do, when in what must have been a typical midcentury scene (1954) my parents called me into the living room to reveal a "wonderful" surprise. They beckoned me toward a large wooden box with a small, dark-green screen—a television set! I'd seen them before, triumphantly displayed in shop windows, flashing moving black-and-white images with a peculiar silver-blue glow. Television seemed interesting enough, but there were so many other things to enjoy and explore. The glowing boxes barely captured my attention. They were dwarfed by the enormity of my imagination. They just didn't seem that important.

My father lifted me up and placed me on my mother's lap, the three of us forming a triangle with the new television. Both parents pointed to the big on/off knob and encouraged me to twist it. Like most young children, knobs and switches intrigued me, I was always ready to open and close things, to explore moving parts, to be further initiated into the world of cause and effect. Somehow this knob was different. It was more mysterious than the others and I was not so anxious to touch it. My parents gently persuaded me. "Go ahead, Mitchie boy. Turn it on." I placed my small hand on the large knob. I slowly fondled it, noticing how it was stiff, smooth, and cool. It took some effort to move it. Finally, I summoned my will and gave the knob a resounding twist until I heard the now familiar demonstrative click.

Whoosh! There were sharp, staccato crackles of sound, a percussive static like the rustling of metallic leaves scattered by the wind. There was a brief flash of light as if my picture were being taken by the television itself. I imagined that the television was pleased to awake from its dormancy, like a bear leaving its cave. It seemed to have a life of its own. Darkness and silence replaced this flurry of activity, until I heard a barely perceptible, low-pitched hum, a noise that reverberated throughout my entire body. And at last, the screen came to life again. As if out of a vacuum, chaotic patterns of black and white danced across the screen before finally coalescing into a discernible picture.

What power I had unleashed! I witnessed a form of magic way beyond what I could understand or had previously experienced. Turn a knob, listen to some whooshes and crackles, and then watch moving pictures. Turn another knob, and see those pictures change—cartoons, ball games, people talking about important things, singers, all kinds of activities and entertainment. Where were all of those people coming from? Were they really speaking to me? Did they live in the box?

Isn't it significant that of all my early childhood experiences and my otherwise dim recollection of those events almost fifty years ago, a memory that stands out is this first encounter with television? I am intrigued by this memory, whether real or partly imagined, because in unfolding the details of the encounter, one which seems very real to me, I unravel some of my deepest impressions of television and re-create the "beginner's eyes" that help me place its perceptual impact in perspective. Further, I am sure that it was the television that ultimately prepared me for the computer and the Internet, and in describing this incident, I can better understand how these technologies in tandem have forever changed the way I perceive the natural world. The television taught me how to view rapidly moving images, and got me accustomed to scanning wide swaths of information in short periods of time.

The first time I turned on the television set I knew I had done something totally awesome. I was "in awe" of what lay before me, amazed at the worlds of information at my disposal. It made me curious, interested to explore its spaces, and to discover what it made possible. I feel the same way today about the Internet, delighted by the prospect of zipping through information, to see what else lies out there, to discover a new website as a new source of learning. For me, the Internet holds out the promise of learning more about the world, and I enjoy surfing the Net just as I love exploring books, or walking through the forest.

But I never watch television or surf the Net unencumbered. Both activities engender a subtle anxiety. After an hour or two of viewing, a kind of myopia sets in. I feel hyper and unsettled, with no sense of ground, having moved through so many images so quickly. All of my senses are literally channeled into a very specific way of viewing the world. I feel trapped and surrounded, finally compelled to look away from the screen, and recenter my focus. Yet after a reasonable break, I am usually ready for more. I am anxious to dive back in and see what else I can find. The researcher in me is compelled to fully survey and explore uncharted territory. Originally the television and now the Internet tantalize and seduce me by holding out the prospects of worlds unseen and information untapped, but they also make me extremely wary and I wonder whether the learning that comes with their use entails a steep perceptual sacrifice.

When I teach courses about perceiving global environmental change, I ask my students to construct similar *technology memoirs*, to remember their first encounters with controversial and influential technologies like television, cars, computers, or airplanes. These memoirs have great instructional merit as a means of better understanding how these technologies influence perception. There is much discussion of the extraordinary pace at which technological "advances" change habits and routines, yet how often do we afford these changes the reflective attention they deserve?[2]

I ask students to gather their impressions about these controversial technologies. It's evident that they are deeply ambivalent, simultaneously condemning their use and promoting their virtues, while becoming habituated to their use. I encourage my students to use autobiographical experience to investigate the source of their ambivalence, and to do so in a specific way. How do technologies of speed, distance and information influence your experience of globality (see chapter 2), how you view ecology and natural history, and how you gather information. How do they influence your pace? How do they make you feel?

I suggest to students that they make a list of the "perceptual technologies" which they most frequently use and consider which have the most dramatic impact on how they observe nature. I ask them to construct a personal history of their relationship to the most poignant machines, including their first encounters, whether their use has increased or decreased, and why that may be so. They place careful attention on how the machine serves as a perceptual aid, what senses it enhances and diminishes, whether it inspires or frustrates. They note whether there are certain categories of machines that they find more or less appealing. I ask them to remember the precise moment they were introduced to the machine, providing as much detail as they can about the context of the encounter.

Another useful instructional approach is to engage in a *technology fast*. Spend a week without your computer, telephone, television, car, or for that matter, don't use any electricity for a while. Keep a journal describing how your observational faculties adjust to the change. What do you take more notice of? Do you observe local natural history differently? Such an activity may demonstrate the extent to which you take various technologies for granted, how accustomed you become to their perceptual support. Imagine the place where you currently live before it was wired for electricity, before there were roads for cars, before television antennas, cables, and radio waves. Whenever I read a nineteenth-century novel, I am delighted to enter into the unhurried pace of the novelist, to read work that was written under a different technological, and hence perceptual regime.

It's very useful to compare "perceptual regimes" across generations. My grandparents (Russian immigrants) grew up without cars, radio, television, computers, or airplanes but were exposed to all of them at different stages of their lifecycle development. For people of my parents' generation, radios and cars transformed the world. It's interesting to speak with them regarding their first impressions of these machines. For me, a computer is still a technological marvel—I'm astonished to think that I can type these words on a laptop computer, erase them and manipulate them electronically, and send them to an editor via e-mail. My children were born into a world of computers, and for them going online is no more remarkable than the daily delivery of the newspaper.

In constructing my technology memoirs, I realize my attraction for machines that receive, manipulate, and interpret information. As a child I would tune the AM radio dial at night, amazed that I would receive signals from all over the East Coast. I listened to weather reports from distant towns and cities. I tuned in to dozens of major- and minor-league sporting events. I heard dialects and different styles of music. I was fascinated that anyone, anywhere could pick up these same signals and eavesdrop on so many different kinds of activities. My radio was a bedside companion, and from my pillow I would wander long distances via radio waves. Imagine my delight when I discovered an old Hallicrafters shortwave radio in the attic. This radio moved me beyond the East Coast to Europe, Latin America, Africa, and Asia, and in so doing unveiled new secrets and discoveries.

I enjoyed a similar sense of discovery when I first learned of the possibilities of the Internet. I found surfing the Net was similar to browsing in a bookstore. Of course the texture of the experience is much different and I do not want to diminish that difference. What links these activities is the sheer joy I gain from wandering fields of information. I remain undaunted by the speed at which the Internet traverses these fields, always desiring faster chips and download times. Similarly, I like to see every section of the bookstore before I settle down to seriously browse in my chosen corner. There are issues of pace and presence here and I address them later in the chapter. What I wish to convey is that I have always been comfortable with scanning broad fields of information. I delight and marvel in their presence.

Yet, when it comes to transportation, I have always been tentative, if not downright frightened of speed. I cried hysterically during my first speedboat ride on Lake Winipisaukee, New Hampshire. I would never go on the more adventurous rides at amusement parks, being happier in the arcades, or on the slow-moving kiddie trains. For years I wouldn't fly in airplanes. I am prone to motion sickness. I once had to leave an IMAX theater presentation because the simulation made me so woozy. It takes me days to recover from transcontinental plane trips. The combination of speed and distance batters me perceptually, throwing my mind and body out of balance, whereas I find the integration of speed and information enriching and intriguing.

Technologies of speed, distance, and information change your pace, and in so doing, they dramatically alter perception. You can't understand how people interpret global environmental change unless you consider the perceptual impact of these technologies. Is there a difference between a child spending several hours in front of a Nintendo game and an environmental activist using e-mail for the same amount of time to organize a legislative campaign? Surely the content is different, but the perceptual impact may be the same. I do not wish to levy judgment on either activity. Rather I urge that we pay closer attention to their perceptual impact, remaining open-minded to their virtues, and vigilant to their abuse. By observing how they have changed your perceptual view over time, how they simultaneously enhance and limit the senses, you are more capable of assessing their impact. Let's apply this reflective approach by considering how different speeds of movement (paces) reveal different aspects of biodiversity.

Place-Based Paces

What does a good naturalist see while gazing out the airplane window on a plane trip from New England to Costa Rica? Dan L. Perlman and Glenn Adelson, in their excellent text, *Biodiversity: Exploring Values and Priorities in Conservation,* lay out an interesting scheme for teaching about biodiversity. They link pace to perception. You see different aspects of an ecosystem, depending on what speed you're traveling. By comparing the perspective of different paces, you can map diversity across a geographic landscape.

At a "flying pace," you view the biome scale. When you travel great distances in an airplane, you can see how vegetation patterns shift according to climate and landforms. Watersheds are prominent. You can follow a river for hundreds of miles, noting how its drainage determines settlement patterns. You can observe the primary vegetation regimes, noting their continuity and disruptions, using these clues to interpret the relationship between mountain systems and rainfall. Whenever I travel across North America by air, I bring Robert Bailey's *Description of the Ecoregions of the United States* or any suitable physiography text. These aids provide the context for landform, climate, and vegetation relationships. Flying allows you to observe the diversity of biomes across the geographic landscape.

The "flying pace" also provides a broad overview of weather systems, allowing you to observe fronts and air masses, seeing how they cover entire biomes. While flying, you pass under, around, and through air masses that may stretch from the Arctic to the tropics. You gain both horizontal and vertical perspectives on weather systems, seeing how they extend over the landscape, while gaining a sense of their density and permeability.

While flying, you gain perspective on the scope and breadth of human activity. Barry Lopez describes this well in his essay "Flight":

An oceanic expanse of pre-dawn gray white below obscures a checkered grid of Saskatchewan, a snow plain nicked by the dark, unruly lines of woody swales. One might imagine that little is to be seen from a plane at night, but above the clouds the Milky Way is a dense, blazing arch. A full moon often lights the planet freshly, and patterns of human culture, artificially lit, are striking in ways not visible in daylight. One evening I saw the distinctive glows of Bhiwani, Rohtak, Ghaziabad, and a dozen other cities around Delhi diffused like spiral galaxies in a continuous deck of stratus clouds far below us. In Algeria and on the Asian steppes, wind-whipped pennants of gas flared. The jungle burned in incandescent spots on peninsular Malaysia and in southern Brazil. One clear evening at 20,000 feet over Manhattan, I could see, it seemed, every streetlight halfway to the end of Long Island, as far east as Port Jefferson. A summer lightning bolt once unexpectedly revealed thousands of bright dots on the ink- black veld of the northern Transvaal: sheep. Another night off the eastern coast of Korea, I arose from a nap to see a tight throw of the brightest lights I'd ever observed. I thought we were low over a city until I glanced at the horizon and saw the pallid glow of coastal towns between Yongdok and Samch'ok. The lights directly below, brilliant as magnesium flares, were those of a South Korean fishing fleet.[3]

Sadly, airplanes are built to pack in as many passengers as possible, and good window views are rare. See all the heads buried in laptops, books, or portable stereos. You're lucky if the pilot points out a famous landmark. Many people are transported great distances, arriving at their destination without an inkling as to the ecological territory they've covered. This is an untapped educational venue. Is it too naive or utopian to wish that as part of their training, all airline pilots and flight attendants might attend brief courses in environmental interpretation from an airplane?

The "driving pace" corresponds to ecosystem scale. In a car, you can discern forest patches, shrub layers, wetlands, and large animals or birds. It's a good speed for interpreting variations in elevation, climate, soil types, and hydrology, and for detecting patterns of disturbance. Perlman and Adelson note how in the Costa Rican rainforest you can drive past a distinct ecological community each minute (with rise and fall of elevation), whereas "you could drive for hours across tundra or temperate steppe grasslands without seeing more than two or three distinct communities."[4]

The driving pace is ideal for viewing biogeographic patterns of vegetation. It's easy to identify forest types, climate regimes, and the most distinctive vegetation. I like ecosystem field guides such as Kricher's handbooks for eastern and western forests as conceptual aids for the driving pace. By reading them in advance of your trip, you can trace the diversity of forest communities you're likely to encounter on a lengthy automobile trip. You can get a great biogeography lesson by driving several hundred miles north or south during spring or fall. Here, color reveals broad patterns of vegetation, either with the blooming of shrubs and trees or the turning of the leaves.

At a "walking pace," you observe the species composition of an ecosystem, or what Perlman and Adelson describe as the "dominant species scale." Whereas in a car you detect the boundaries and mixtures of biogeographic regimes, when you walk you gain much more information about plant associations, floral-faunal relationships, and niche-habitat arrangements. Walking provides information about indicator species. As I walk through the northern hardwood forest, I see yellow birch, sugar maple, American beech, and eastern hemlock. On even closer observation, as I pass through northern and southern exposures, highlands and lowlands, and various wetland habitats (beaver ponds, swamps, streambeds) I notice subtle changes in the species composition, realizing how diversity is a function of climate and elevation. The northern hardwood forest has many variations, even within a ten-mile radius.

Walking through a forest in Florida, one observes such variations for every foot of elevation!

At the "crawling pace," or the small species scale, the true richness of species diversity emerges. Now you can look carefully at herbs and shrubs, soil flora and fauna, insect lives, bird's nests, frog's eggs—whatever is visible with the naked eye. Perlman and Adelson note how at these scales their students "observe variation and diversity that are a result of microhabitat gradients."[5] On your hands and knees, you can notice the variety of lichens on a single boulder, the insects that crawl on the lichens, the patches of moss that struggle to take hold. In tropical environments, where Perlman and Adelson take their students, microlandscapes teem with the diversity of life. At a crawl, they are best able to point out the subtle variations that distinguish the individuals of a single species.

At microscopic levels, there is the "electrophoresis pace," or the genetic differences scale. "Samples of DNA and proteins are placed in weak electronic fields and under the influence of these fields, the samples spread out revealing genetic strains."[6] An entirely new spectrum of diversity is revealed. The microscope is the best way to study five kingdoms taxonomy (see chapter 5), and to appreciate the life that is only visible using microscopic instrumentation.

Perlman and Adelson not only describe what can be observed at these various paces but they emphasize the importance of slowing down as a technique for appreciating biodiversity. "A key point here is not whether one sees additional diversity as the pace of travel slows—since of course one would expect to see more detail at a slower pace—but rather that entirely new categories of diversity become visible as the pace decreases."[7]

The message is simple—to observe the complexity of biodiversity you have to slow down! The more still you become, the more life avails itself. Each pace reveals some patterns and conceals others. Taken together as perceptual tools, various paces allow for the juxtaposition of scale. A clear-cut looks very different from thirty thousand feet than it appears as you walk the land. An enormous lake may be very beautiful from an airplane, but you can't really tell whether it's clean or polluted unless you walk its shores and sample its water.

The speed of flying allows you to span great distances and observe global environmental change at the biome scale. The flying pace reveals broad patterns of global landscape, atmospheric weather systems, and human geomorphological agency. But it conceals biodiversity. It's very difficult to see life forms from the air. There are far more people riding airplanes than there are crawling around on their hands and knees, exploring microhabitats and sampling soil fauna.

Crawling, walking, running, bicycle riding—these are place-based paces that enable you to observe the living community of landscape. The high speed of cars, trains, and planes (driving and flying paces) necessarily protect you from too much contact with the elements. You are going so fast that you must be contained. This enclosure serves as a protective barrier that not only insures safety but removes your ability to observe nature with all of your senses. At high speeds, your observations become more cognitive and symbolic. The speed of the vehicle overwhelms the visceral contact with nature. You're looking at the world behind glass. The patterns that you perceive at these faster paces, while traveling at high speeds, are made tangible in relationship to your place-based, earth-paced contact.

I find that when I travel by airplane, the shock of such a fast speed is so frightening and exalting that I require the stability of some kind of conceptual hold. Like a meditation process, or even a form of mantra, I remind myself that I am observing the biosphere at a flying pace, and that it's a temporary view serving to expand my vistas until I can return to the place-based paces to which I've grown so accustomed. The challenge is in learning how to attain the deliberate gaze of the naturalist even when traveling at high speeds. It's important to remember what you're looking for and why. This is also true when using computers to perceive the biosphere. In the next section, we'll see why a place-based pace is essential for navigating the Internet.

www.biosphere.edu

While flipping through the sumptuous pages of the National Geographic *Satellite Atlas of the World* you might notice a two-page map of the world labeled "Pulse of the Planet." It portrays the "living planet," a composite picture of "tens of thousands of mosaiced satellite images" depicting the earth's "productivity." Ocean areas rich in phytoplankton are purple and land areas with "high-potential plant productivity" are portrayed in green. Framing this colorful spread, laid out like hors d'oeuvres on a plate, is a semicircle of small oval global portraits, each using color to depict the global ranges of biospheric dynamics—topography, cloud amount, precipitation, snow depth and sea ice, day/night surface temperatures, distance, wind speed, surface waves, sea level variability, and sea surface temperatures.

Note the descriptive text: "Measuring, monitoring, and modeling these myriad elements—and their effects on the biosphere—from space with regularity and certainty enable scientists to better understand and quantify systems interrelationships and the role of ocean biology in the global carbon cycle, key to the Earth's survivability. Far from static, these global pictures are constantly changing. As CAT scans to the human body, they represent a new vision of the earth's vital data as it lives and breathes."[8]

You can spend a long time studying this page. It contains an enormous amount of interesting information and it is strikingly beautiful. There are many intriguing patterns to observe as you contemplate the flow and movement of ocean and land. As you get more deeply engrossed in the map, the illustrative colors seem to vibrate and it is easy to imagine the pulse of the planet.

The remainder of the atlas is equally breathtaking, a work of art on behalf of global change science. There are remarkable pictures of the earth from above, depicting topography, watercourses, geological features, and the heavy hand of human impact. There's a stunning portrait of the United States, a cloud-free topographic view, affording outstanding resolution of the nation's physiography. The caption describes how "this natural-color mosaic pieces together 432 different images to achieve its cloud-free clarity."[9] On close inspection of almost all of the maps in this atlas, you realize that every picture is rearranged, filtered, manipulated, patched, and squeezed to attain its appropriate resolution. This is an atlas entirely generated by satellite photographs and their mediation through computer imagery and human design. Through this cartographic process, you ascertain many patterns of global environmental change. Surely this is the epitome of a "geographic information system." Indeed, the atlas editors frequently refer to how computers and satellites have transformed cartography, providing data collection systems and monitoring capabilities with powerful diagnostic ability, all in the service of global change science.[10]

Stephen Hall, in *Mapping the Next Millennium,* shows how the convergence of satellites and computers contributes to profound biospheric discoveries. Paleoclimate maps and computer models of atmospheric change generate the global warming hypothesis. The development of a TOMS map (*t*otal *o*zone *m*apping *s*pectrometer) helped uncover the hole in the ozone layer. In *Land Mosaics,* Richard T. T. Forman describes how geographic information systems and images from satellites have "revolutionized our perception and approaches to understanding landscapes and regions,"[11] serving land-use planning, biodiversity studies, and biogeography. In *Cycles of Life,* Vaclav Smil shows how computer models and satellite maps reveal changes in the carbon cycle, allowing scientists to trace energy fluxes and biogeochemical flows.

Hall describes the significance of this extraordinary mapping technology:

The real breakthrough—messy and beyond category, a chain reaction rather than a single explosion—is in twentieth-century science's ability to measure, and therefore to map, a breathtaking range of spatial domains. Scientists, our latter-day explorers, are charting worlds that Magellan and Columbus, in their most homesick and delirious moments, could never have imagined. With stunningly precise new instruments of measurement developed over the last half century and with the tremendous graphic power provided by computers over the last two decades, everyone from archaeologists to zoologists has been able to discover, explore, chart, and visualize physical domains so remote and fantastic that the effort involves nothing less than the reinvention of the idiom of geography.[12]

Hall then suggests how "the widespread availability of computers with specialized graphics software puts the equivalent of a cartographer inside every computer."[13] Glance at the last several pages of the *Satellite Atlas of the World* and you'll note the acknowledgments to the various agencies, companies, organizations, and universities that assisted in its planning and preparation. There are seventy-five listings, and most remarkable is how each listing has a corresponding website. In effect, each website is an atlas of its own. These vital biospheric data, these glorious maps, are available to anyone who has a computer and uses the Internet. If you want to bring the biosphere home by virtue of computer-generated maps and images, it is very easy to do so.

What is one to make of this extraordinary opportunity and how might it be navigated? First, a pronounced caution. Remember that a point of view is contained in any map. Our satellite atlas is a testimony to symbolic mapmaking, but all of the maps are contrived and both their logic and resolution are ripe with interpretation and controversy. "A map," Stephen Hall reminds us, "above all, is a worldview committed to paper."[14] These maps reveal resources to be exploited as well as ecosystems to be saved. Richard T. T. Forman advises us to beware the overlay as "the additive effect of errors in each information level . . . limits the accuracy and interpretation of a final integrated overlay image." This has crucial design implications, too. "And when presented with a sequence of multi-colored images of a changing two- or three-dimensional mosaic, the human mind cannot follow and understand most of the changes; thus a presenter's message is highlighted and the evidence is obfuscated."[15]

These wise observations translate well to the Internet, where biospheric data may move too quickly for anyone to reasonably assimilate them. It is fair enough warning to behold a map replete with beautiful colors and images only to discover that the patterns depicted require interpretation and discussion. Beauty is in the eye of the beholder. The speed and abundance of computer-generated data don't erase the need for deliberation and moral choice.

Speed is of the essence on the Internet, and there are times when your mind moves so quickly that you feel as if you are literally flying through images and websites. You are willing to take as much as you can get— cruising through virtual information as fast as a jet flies over the landscape. It's exalting to scan and synthesize and surely this is what the Internet allows you to do. "The world henceforth will be run by synthesizers," says Edward O. Wilson, "people able to put together the right information at the right time, think critically about it, and make important choices wisely."[16] The "wise choice" requires you to sit back and reflect on what you have seen.

A second caution—don't let the screen images move too quickly. Find ways to let your body catch up with the pace of information. Return to the five senses! During breaks on the Weather Channel, the time between the assured smile of the forecaster and the relentless pitch of the advertisers, there is a respite from talk. They broadcast the latest satellite views of North America, but do so in a time-lapse framework. So for a few seconds you can observe what amounts to a day's worth of weather. Watch the cloud formations move over your region. Before you've even had a chance to digest the scene, you move onto another section of the country. It takes about one minute to present these sequences for the entire nation. Although I've learned how to watch these images to get the information I need, I am always behind their pace, wishing they would spend some more time in New England so I can trace the nuances of cloud cover and movement. But just as I've picked up a pattern or two, we move to the Southeast. It seems as if I'm always a thought moment behind. I look out the window and observe the sky. I wander outside and feel the wind on my face. I watch the satellite photos again, grounded in the tangible reality of my sensory impressions.

The speed of the Internet presents a similar dilemma. Hypertext is a form of overlay—a means of travel in a turbulent sea. The winds and currents of your browser flow and sway across great distances. Without a navigational theme and steady ballast, it's easy to get seasick on the waves of information. Home port is terra firma. You must always know where you come from, where you are going to, and why you're traveling there. It's just too easy to get lost. What is your reason and purpose for surfing the Internet? What urgency motivates you to download information at such breakneck speed? And what is concealed at your fast pace?

If you use the Internet to explore the biosphere, let your solid ground be the visceral observations of backyard ecology. Then you'll have a place to which you can safely return. Reestablish place-based paces. You'll know what you're trying to achieve, and you can take side trips along the way—juxtapositions of scale, long-distance wanderings at websites very far from home—returning with the data that you need.

And what reams of data lie waiting to be explored! An hour of cursory research will reveal numerous global change websites yielding dozens of links providing satellite imagery, remote sensing data, and field-based natural history observations, with full explanations of the relevant research techniques and modeling approaches. Start at the Global Change Master Directory (http://gcmd.gsfc.nasa.gov) for a comprehensive list of data sources, indexed by scientific topics, regions, ecosystems, and species. At the NASA GCMD Learning Center there's a link called "data you can use" offering data sets available for download and analysis, covering various aspects of global change and the Earth system. Here are a few examples. Follow the ozone data from the TOMS instruments (referred to earlier). Assess global warming by checking out the Global and Regional Cloud Cover data from the International Satellite Cloud Climatology Project. Witness biogeochemical cycling with the carbon dioxide record from the Mauna Loa, Hawaii observatory. Other NASA sites provide galleries and photo libraries of atmospheric, earth science, and oceanographic photos. Many of these sites provide a K–12 science curriculum, networks for teachers and students, and specific guidelines for how to use the data or to generate your own.

Equally impressive are the online, field-based natural history sites. Journey North (www.learner.org/jnorth), funded by the Annenberg Foundation, enlists hundreds of teachers and thousands of students in efforts to trace the migration of the monarch butterfly. At their website you see a map of North America, with dozens of small dots corresponding to the latest sightings of the butterflies as they seasonally migrate north or south. Click on the dot and the name and location of the observer pops up. You can print out a monarch butterfly migration checklist which helps you to structure your observations, listing categories such as the date and time of sighting, weather conditions, the activity of the monarch, the condition of its wings, and the height of the milkweed and other local blooming flowers. At another link teachers can search for partnership classrooms so their students can exchange data with kids from around North America. Journey North also has a phenology data exchange in which students submit their observations of spring, tracing its northward progress with partner schools as they share their data on the Internet.

The Cornell Laboratory of Ornithology (http://birds.cornell.edu) has a website listing their "citizen science" projects. They enlist birdwatchers in various data-gathering projects. With Project FeederWatch, observers count the birds that visit their feeders from November to March, sharing their observations with field researchers. The Birds in Forested Landscape program asks experienced birders and professional biologists to "help collect data that will be used to determine the effects of fragmentation of North American forest birds." Other projects include the Cerulean Warbler Atlas Project, Classroom BirdWatch, the House Finch Disease Survey, and Project PigeonWatch. "From backyards to remote forests, these citizens represent the world's largest research team. We call them citizen scientists."[17]

The Internet is a remarkable tool for barefoot global change science (see chapter 5). We now have a sophisticated, international network for the gathering and dissemination of global environmental change data. These collaborative online data banks allow you to download or upload "live" in the field observations, comparing what you see with information gathered from thousands of other observers. In a way, your five senses serve as antennae for a virtual and visceral monitoring system. When you upload a monarch butterfly sighting, your data move from your eyes and ears to a global network! The computer serves as the portal for translation and exchange of information.

This instantaneous reporting serves a sentinel function. It's much easier to understand the changes you observe when they are placed in a broader, global context. From your local field site, you can broadcast the ecological news of the day. This information may be disseminated through a formal news service, or other informal messaging formats (listservers, conferences, e-mail, and other networks). You can subscribe to an environmental news network to scan the global wire services. You can create your own mailing list to get a word out to a targeted audience.

Or you can combine both formats. Word spreads fast. In the political arena, there is much ecological mileage to be gained from the swift-moving currents of global data exchange.

The "grassroots" metaphor is fitting in the face of such complex networks. Grassroots implies keeping things in the field, from the place where you come from. Citizen participation relies on the experiences and stories of the people, species and landscapes that are close at hand. Hands-on observational research is the bedrock upon which these far-flung communications rest. It takes a heterogeneous community of dedicated, experienced observers to interpret the information, to make wise choices in that regard.

We all know how what appears as state of the art on the Internet today seems dated and tired tomorrow. Surely this data pool and its virtual and visceral linkages will proliferate, spurred by faster speeds and more vibrant colors. Access to global change data will become more transparent and available. However, it provides me with no relief knowing that from the corner of my study, with my laptop perched on my thighs, I can download attractive images of the earth from space, or vegetation maps of my region, if the loss of biodiversity continues unabated. But it is very reassuring to know that from my outpost, I can share whatever observations cross my path, and that there are like-minded observers engaged in similar activities—sharing, teaching, exchanging, and learning together.

The emergence of a coherent, ecosystem monitoring function—the blending of the Internet, GIS systems, satellite images, and field-based natural history research—is an indispensable tool for interpreting global environmental change. And yet, the inexorable pace of Internet participation dramatically changes how nature is observed, accelerating the pace of information, creating expectations for speed, and perhaps, placing a greater conceptual distance between the object of study (the ecosystem) and your visceral understanding of it.

Many people experience jetlag when they fly across biomes and time zones, requiring a period of psychological and physiological adjustment. I view this as the shake, rattle, and roll of high-speed transportation, jolting you out of the place-based pace to which your body is attuned. A high-speed Internet chase can have a similar effect, a sort of *netlag* if you will, with attendant feelings of vertigo, groundlessness, and disproportion. Both forms of high-speed travel can be simultaneously exalting and overwhelming. And they both share a particularly insidious quality—the quicker you move, the faster you expect to travel next time. Both networks offer a "rush," and in such a state of mind, what is more irritating than a bottleneck, whether it's waiting for your plane to take off, or a website to download its images?

The Internet presents the biosphere at breakneck speed, in sharp contrast to the deliberate gaze of the place-based naturalist. Ironically, the very urgency of global environmental problems propels the very state of mind that you may try so hard to resist. Time is short, and we must act now—here is the call of both the activist and the advertiser. The Internet is too powerful a conceptual network to shun, yet too dramatic a perceptual tool to unquestioningly embrace. Speed kills! A middle path is required, one based on a deeper, place-based sensibility. What is the necessary balance between the virtual and the visceral and how might it be attained?

Perceptual Checks and Balances

Cultivating biospheric perception is a learning practice and if the Internet and interstate are inextricably built into this practice by virtue of their daily use, then appropriate interpretive guidelines are surely warranted. These guidelines are informed by place-based perceptual ecology. They are a means of slowing down when you find yourself traveling too quickly, of using deliberation to measure the instantaneous, and of reflecting on the purpose of your journey.

Learning how to use the Internet and interstate as approaches to biospheric perception requires skill and practice. When you go birding for the first time, it takes some practice to use binoculars well. With experience you learn how to adjust the focus depending on what you're looking at. There is a pace to using binoculars that matches the pace of birdwatching. The same is true of the microscope, telescope, or any instrument of magnification. So it is with any powerful perceptual instrument—tools of information, speed, and distance included.

With the Internet and interstate, a primary guiding principle is to *modulate pace.* Always know how fast you're traveling and how your speed influences what you perceive. When flying, use the opportunity to think at the biome scale, ready to observe global environmental change across entire landscapes and regions. When surfing the Net, ride the information waves to rapidly scan biospheric data. But don't forget to slow down when you hit the trail. Otherwise you'll never see what's in front of you. The depth is in the details.

Modulate accordingly. What rhythms of landscape emerge at your traveling speed? Modulation implies attunement, an adjustment to rhythm, a making of melody, a search for the right key. You sing in a key that fits your register. Perhaps every pace has a key and rhythm that's specifically attuned to the tone of your voice. You may find that certain paces demand specific rhythms.

Modulating pace provides access to multiple scales. As you get comfortable with various paces, it becomes easier to compare scales. With awareness of pace, you can move from the biome to the species level and back again, comparing your impressions, finding multiple views for gazing at the biosphere.

Remember, too, that pace structures perception, and what you see is influenced by how fast you move. *Trace what is revealed and concealed* at every scale. If you're a high-speed globetrotter who never leaves the Internet or interstate, you will only view the world from a fast-paced perspective. If you only see the world from space, you'll discover some exhilarating biospheric patterns, but you'll never know what's going on down below. If you view nature exclusively through the Internet, you'll have wonderful data and images to explore, but your visceral connections may atrophy and your knowledge of ecology will be forever abstract. Any window only lets in what its dimensions allow, framing your view according to the resolution of the instrument. The screen is not the territory!

Place-based paces provide the advantage of detail and deliberation. There are some things that you'll never see unless you slow down long enough to allow them to happen. It's not so boring to watch the grass grow because there's so much more to observe while you're waiting. Remember Perlman and Adelson's observation that as you slow down, more aspects of biodiversity are revealed. Patience reveals, rushing conceals. With patience comes elucidation. Clarity requires context and experience. Patience and clarity yield insight.

An instrument will only amplify and shape what you already know. When used appropriately, it will display new dimensions and appearances—magnifying or shrinking, broadcasting or interpolating. But instruments bring you no closer to the biosphere than you already are. Beware the illusion of the intimacy garnered through perceptual aids. That is why it's so crucial *to balance the virtual and the visceral.*

Consider the tangible impressions you gain when you walk the land with full visceral (bodily) attention. Contrast this with a photograph of the same land taken from an airplane. Or from a painting in a museum. Ponder the stories attached to each experience. Now read Barry Lopez:

If I were now to visit another country, I would ask my local companion, before I saw any museum or library, to walk me in the country of his or her youth, to tell me the names of things and how traditionally, they have been fitted together in a community. I would ask for the stories, the voice of memory over the land. I would ask to taste the wild nuts and fruits, to see their fishing lures, their bouquets, their fences. I would ask about the history of storms there, the age of the trees, the winter colors of the hills. Only then would I ask to see the museums. I would want first the sense of a real place, to know that I was not inhabiting an idea. I would want to know the lay of the land first, the real geography, and take some measure of the love of it in my companion before I stood before the paintings or read works of scholarship. I would want to have something real and remembered against which I might hope to measure their truth.[18]

For every hour you spend in front of the computer, how much time do you spend gazing at the sky and watching the clouds? For every hour you're in the car, how much time do you spend walking the land? What ratio do you set for yourself and by what criteria is it set? This is a fundamental issue for place-based perceptual ecology. Your answer will tell you much about the learning spaces of your life, and the means at your disposal for interpreting the biosphere. I offer no set formula. I urge a fair balance, one informed by your deepest values. For myself, I know how I feel when the proportion is out of kilter. I'm aware how easy it is to lose touch with wild nature, or to interpret the biosphere exclusively through a looking glass.

An interesting way to balance the virtual and visceral is to rethink your perceptual application of technology. Video artist Paul Ryan explains how you can use a video camera in an active and participatory way so as to "push the envelope of perception."[19] He developed a "handheld, continuous camera style based on flowing t'ai chi movements," enabling him to "meditate in motion through the camera."[20]

Ryan describes how in 1973 he spent a year living by a broad, rocky streambed, studying water flow patterns. To unlearn his adult perceptual habits and "return to an innocent childlike state of mind," he would strap a camera to his head and "crawl around in the stream like an infant, letting the different water flow patterns into my brain." This experience led him to develop "a vocabulary of different patterns of water flow."[21] For the next twenty years, Ryan recorded water flow patterns at waterfalls along rocky coasts, and over the open ocean. "Seeing through the camera, and being able to replay what I see at varying speeds and in both directions, has given me an understanding of water that I could not have gotten with the naked eye."[22] So inspired, Ryan developed an Earthscore Notational System, various means of using video observations to record ecosystem processes. His proposed "ecochannel" uses participatory video, satellite imagery, and remote sensing as a means of building community awareness, enabling the users of these technologies to simultaneously explore perception while recording and interpreting an ecosystem.

This is a fine antidote to philosopher David Rothenberg's observation that "the more we learn about how to use an instrument, the less we think about it as we use it."[23] With the habitual use of any technology, you grow accustomed to seeing the world exclusively with its aid. Hence Rothenberg's sage advice. "Whenever one adapts a new technique, it is essential to remember the way things were before it came to be available."[24] Reflect on where you've come from. Remember how you used to view the world.

The most effective perceptual "checks and balances" tool is in knowing how to *engage the on/off switch.* Is your computer always on? Do you carry a cell phone everywhere? Do you drive to every place you have to go? It's so convenient to remain habitually connected to the interstate and Internet. It's much harder to disengage from those networks. Disengagement removes you from the great web of human commerce and discourse. A good way to leave the networks is to shut them down for a while and see what emerges in their place.

There is great wisdom in knowing the right time to use an instrument. The finest improvisational musicians understand when to remain quiet. Sparse notes may have as much impact as flashy runs. The most powerful musical passages reflect pause, rest, and measure. It's the silence between the notes that nourishes the music. Biospheric perception relies as much on moments of stillness and memory as it does on snazzy instrumentation. One measure of both good art and science is the extent to which one's insight and depth find a home in an instrument. The on/off switch is a fine tool for exploring the spaces in between amplification and reflection. Rothenberg reminds us that "technique does not replace awe, and the universe remains more than we make of it, or what images we supply to explain it."[25]

A Delicate Balance

Gary Snyder, in his stirring visionary manifesto "Four Changes" (written in 1969) referred to "computer technicians who run the plant part of the year and walk along with the elk in their migrations during the rest."[26] Now, over three decades later, there are far more computer programmers than there are elk, and few of them can leave their screens long enough to fulfill this Pleistocene destiny. As the world gets wired, the wild gets left behind. There are plenty of technicians, programmers, and entrepreneurs, who spend all of their waking hours designing for speed and efficiency, who embrace the Internet and interstate with almost religious zeal. If only there were as many people devoted to conservation, studying the biosphere and its diverse life forms! In this regard, the world is out of balance, and the fulcrum is weighted toward a virtual future.

Both computer programmers and conservationists know that the world is always much more than it appears. Just as the conservationist may design a website to promote an educational approach or political agenda, the computer programmer may spend two weeks every year backpacking in a remote wilderness. This is the reality of the twenty-first century and it is one that we all must negotiate. Virtual and visceral worlds interpenetrate and sometimes the boundaries are not readily discerned.

Although I try to reside comfortably in both worlds, I find little more than an uneasy peace, provoked by an unsettling and resilient ambivalence. I am attracted to the computer screen and inspired by the wild. I modulate accordingly—changing speeds to travel distance and survey information, using my place-based proclivities as anchor and sail. I move from the Internet to the forest and back again, from the weather channel to the wind on my face, from the environmental news network to the smells of a morning walk. I take heart in Gary Snyder's vision, hoping that ecological wisdom lies in these divergent, but simultaneous paths.

But I'm not always sure that this modulation works. I wonder if I'm subtly and imperceptibly drawn into a world of speed that outstrips my reflective capacity. Maybe the prospect of cultivating biospheric perception via the Internet is an illusion. Perhaps I have fallen prey to Paul Shepard's warning of "virtually hunting reality in the forests of simulacra."[27] Here lies a resilient ambivalence. I aspire to probe this ambivalence, to use the tension creatively, to watch myself carefully, and never relinquish the deliberate pace which I claim to endorse. I'll worry even more when I abnegate that responsibility, when I'm too comfortable with my rationale. I'd rather reside in both conceptual worlds and take on the perceptual challenge, letting the doubts remain.

I can't endorse this ambivalence without some form of sanctuary. When the pace quickens beyond control, I retreat to the visceral forest where I can more easily attain stillness—moving at a walk or crawl, or just sitting quietly in the presence of the biosphere, perceived unaided, unamplified, without any instruments. In those moments, the pace of my thinking corresponds to the present moment, and with deliberation and reflection the great expanse of the biosphere begins to gradually unfold. I require sanctuary to restore deliberation.

There is reassurance in the province of two preserves—the sanctuary of wildness and the Internet-free zone—places where you can walk the living landscape free of the pulse of electricity, the hum of traffic, and the towers of human babble.

As agents of speed and efficiency the Internet and interstate have no speed limits. They accelerate relentlessly, parallel to the engines of commerce that fuel their use.[28] In this time of unhesitating endorsement, in what manner does one suggest judiciousness and prudence? There is a profound difference between a search for knowledge and the obsessive pursuit of one's own reflection. One path leads to wonder, the other to narcissism. Always ask yourself—what is my purpose in using this network? Just as technologies of information, speed, and distance serve to enhance and diminish biospheric perception, so they strengthen and lessen the human condition.

On the verge of the sixth megaextinction, we face unbounded threats to the diversity of life. The aspiration to cultivate biospheric perception is a response to this condition. A place-based perceptual ecology opens the senses so you can more fully observe the unbounded glory of your ecological and evolutionary heritage. There are so many ways of exploring this heritage—crawling on your knees through a wetland on a humid summer morning, watching warblers on their spring migration, or observing biogeographic patterns on a series of computer overlays—all so many means of learning how to affiliate with life.

Remember Vernadsky's challenge—how the extraordinary dynamism of life is more extensive than you can ever really know. Sometimes it requires stillness and patience to reveal the full motion of life. Other times you need the perspective of speed and distance to divulge the dynamic eddies and whirls of life swirling through the biosphere. These approaches are balanced on the fulcrum of place, so you can always return to the intimacy of home. So, nourished, grounded, and rooted, you can take the plunge into the twenty-first century, taking full advantage of its remarkable instruments, yet keeping both feet planted in the natural history of the place where you live.

Place-Based Transience

A Recurring Dream

I have a recurring dream. I am wandering the streets of a strange yet familiar city, a surrealistic Bronx, resembling the urban landscape of my college years. The sky is gray. A cold, damp mist permeates my body. I know that I have another life somewhere, embedded in a hilly, hardscrabble terrain where my house sits in a soft wooded hollow. I have a wife there, and children. Deep feelings of belonging surge through my body. Images of that life coalesce and disappear, yet I can't seem to find it. I'm not even sure that it ever really existed.

Desperate to alleviate the cold loneliness and confusion, I traverse alleys and plazas, labyrinths of asphalt and brick. I float among waves of pedestrians, searching for the elusive small flat where I lived long ago. I turn each corner hoping to find this temporary place, to set anchor there so I might re-create the past and discover the way home. But every intimation of familiarity dissolves into illusion. Where is my place? Why has it been taken away from me? Is it merely a dream of what might have been, one of dozens of alternative life stories?

Every generation in my family has lived in a different place. It is no surprise that my current home place might seem like an ephemeral security. My dream of dislocation is not only a shadow anxiety, it is at the core of my family history. I can still vividly recall my grandmother's stories of life in the old country, told with a sweet Yiddish accent, reflecting a generation of Jewish immigrants who shared strong cultural bonds despite their geographic wandering. Both sets of my grandparents were born in the Russian pale, the children of marginalized peasant farmers, rabbis, teachers, and shopkeepers who lived in perennial fear of the next Pogrom. They emigrated to New York and the Mediterranean island of Cyprus, respectively, the birthplaces of my parents. Eventually my parents moved to Long Island where I grew up. I left twenty-five years ago for New Hampshire.

My proximity to this experience is similar to the memories of millions of Americans who are only a generation or two removed from Ellis Island or numerous Atlantic or Pacific ports of entry. Africans, Armenians, Chinese, Irish, and dozens of other ethnic groups have a diasporic tradition that entails suffering alongside opportunity. This is an important demographic dynamic—the dislocation that accompanies European economic expansion, transforming habitats and cultures, creating emigrants, exiles, refugees, slaves, traders, and imperialists, an assortment of people in migration by coercion, chance, or choice.

Throughout this book I espouse place-based perceptual ecology as the foundation for learning about global environmental change. Yet few people stay in one place long enough to observe its ecological and cultural subtleties and nuances. Typically one's rootedness is less a function of time spent in a landscape and more a matter of various symbolic affiliations—ethnic and religious bonds, family histories, business ties—so many factors may determine how you construct your heritage. With economic globalization, the movement of peoples is even more pronounced. Whether it's business opportunity, political uprootedness, habitat degradation, or the sheer joy of living in many places, the demographics of transience seemingly render a place-based orientation anachronistic.

In this chapter I consider the terms of a diasporic residency by describing the seemingly contradictory concept of place-based transience. What does it mean to be a resident who is just passing through a landscape, and what are the implications for perceiving global environmental change? How can we use migration and diaspora, among both people and species, as a means to bring a biospheric perspective to local ecology and community? Perhaps transience is an approach to broadening scale and perspective. In the twenty-first century, one may cultivate numerous allegiances—to home bioregions and far-flung places—creating a global community that resembles a wild diasporic garden. Can the idea of transience open minds to an ecological cosmopolitanism based on multiple fidelities and the vitality of both biological and cultural diversity?

The inevitability of ecological transience offers a perceptual opportunity—a means to observe and internalize the diversity of peoples and landscapes. The first half of the chapter opens with a discussion of contemporary human mobility, placing migration in both a historical and ecological perspective. Then I consider the correspondence between human and animal migrations, suggesting that migration is a biospheric process. How do these migrations bear on attachment to place, one's relationship to community, and biospheric perception? And why is an understanding of transience crucial to interpreting global environmental change? The second half of the chapter describes the limitations and potentials of a diasporic residency—how do you live in a place even if you are just passing through? This, too, is a perceptual foundation of biospheric perception, learning how one's place is connected to dozens of others, and how the matrices of those connections weave a cosmopolitan web—the mingling of peoples and species, the confluence of habitat and history, and the convergence of ideas and landscapes.

Human culture has always been on the move, replete with punctuated migrations. Periods of settlement are interspersed with periods of expansion and dispersal. There is a context for transience in human history, whether spurred by a quest for subsistence, overshoot of carrying capacity, or sheer curiosity and wonder. Check out any atlas of world history. Note the flow of movement and color signifying the transformation of landscapes, carving new borders and boundaries. Observe the interplay of ethnicity and geography, forming and dissolving as quickly as you can flip through the pages.

The anthropologist James Clifford (1992) suggests the idea of traveling cultures which consist of "missionaries, converts, literate or educated informants, mixed bloods, translators, government officers, police, merchants, explorers, prospectors, tourists, travelers, ethnographers, pilgrims, servants, entertainers, migrant laborers, recent immigrants, etc."[1] As I trace my ancestry through maps, memories, and stories, I realize that I am a member of this traveling culture. No wonder I am attracted to the moral and ethical mores of place-based environmentalism—an approach to learning and living that cultivates respect for landscape and habitat, community-based decision-making, and mindful inhabitation. These qualities are an emotional and intellectual response to a collective anxiety dream, forged in a crucible of uprootedness amid the turbulence of global environmental change.

I find solace in the stability of my home landscape, believing that with increased awareness of the flora and fauna of this place, I will no longer be a transient, one who just passes through. Via intimacy with the local ecology, I aspire to become native. Yet a quick review of the ecological history of the landscape reveals a series of dramatic transformations. Only eighty years ago, around the time my grandmother stepped off the boat, this forest was farmland and sheep roamed the hills. Even farther back, a Pleistocene history reports a series of habitat changes reflecting the retreat of the glaciers and the northward movement of trees, grasses and animals.

Who came before me? How long have the various residents been here? Which came first—oak, beech, birch, maple, or white pine? What about the vegetables in my garden? New World squash coexists with Old World broccoli. The dandelions that sprout each spring arrived only several hundred years ago, cargo on European tall ships. Various residents have come and gone as the climate and landscape have shifted. Only ten thousand years ago, dire wolves, short-faced bears, and giant beaver roamed this landscape.[2] Not long before that, powerful glaciers carved the hills. And what does a global warming future hold?

There is a context for transience in ecology too, both through spatial and temporal dimensions, and not just at the behest of human impact. Migrating songbirds have winter and summer homes. So do butterflies and bats. Who then is indigenous? Who lays claim to being native? On some level, depending on the scales of ecology and history, we are all transients—humans, oak trees, and mountains alike. My intention is to explore the significance of transience for biospheric learning.

Where then might rootedness reside? How do you come to be in any place? As a diasporic Jew, how did I arrive in the Monadnock region of southwest New Hampshire? And how did the phoebe, who returns each spring to the same nest over a broken lamp on my garage, manage to find its way back home? We are both migrants in our own way, searching for opportunity, survival, a niche in the world. Depending on the scale of observation, we are all transients, just passing through the landscape, making our way through life. The phoebe and I, and many other people and critters too, share this landscape by virtue of the confluence of habitat and history. Often you just stumble into your community and hope that you'll be welcome. You help make yourself welcome through your good work in return.

Road Trip

A good way to think about the breadth of human mobility is to remember and locate all the different places you've lived in. The average American moves every four years, so if you're an American and you're forty years old, chances are you've lived in ten different places. There are a lot of different reasons to move—job opportunities, schooling, family dynamics, geographic preference, getting away from a bad situation, or looking for a better one.[3] For some people, mobility reflects the privilege of their wealth, for others, it's a characteristic of uprootedness, and sometimes a measure of desperation.

Much is made of the particularly mobile quality of life in a global economy. Supporters suggest that it allows for unprecedented economic and political choice. Critics bemoan the instability, the breakdown of community, and ultimately the decline in citizenship and environmental awareness. But supporters and critics alike probably share one thing in common—they undoubtedly travel a lot. I am often amused at the number of times I am hundreds or even thousands of miles away from home, advocating the importance of paying attention to the place where you live.

The relative convenience of travel—the ubiquity of the airplane, the car, and even the bicycle—facilitates commerce and mobility, and, of course, tourism, often touted as the world's largest industry. Indeed, this seamless integration of commerce, mobility, and tourism, is so interlocked as to be a trademark of our time. People think nothing of long-distance commutes, multiple residences, and seasonal employment, and you can find these attitudes at both the high and low end of the economic spectrum.

Recently, the *New York Times* ran a four-part investigative series entitled "Here and There: Immigration Now" which profiled some of the international dimensions of this global mobility. Fernando Mateo, who owns a money-transfer company, commutes between Westchester County and Santo Domingo. "Many a day, he and his wife, Stella, start out in blaring traffic on the Grand Central Parkway and end up on horseback in the verdant Dominican countryside, cantering down to a river to feast on rum and goat." The article asserts that modern immigrants, depending on what they can afford, often straddle two worlds, allowing them to have two homelands. "The population of Chinantla, Mexico is evenly divided between that tiny town and New York City, but the Chinantecans still consider themselves one community—2,500 people here, 2,500 people there. In New York, the first-generation Chinantecan immigrants are waiters, garment workers and mechanics; back home, they are the big shots whose earnings built the town's schools and rebuilt its church."[4]

Hamad Ali, who drives a taxicab in Manhattan, sends much of his paycheck home to Pakistan. He communicates home by periodically using a videophone parlor where he can speak with family members for five dollars a minute. Leonid Slepak, the son of two "refuseniks," fled Russia in 1979. He now has a business that requires him to travel back and forth between Moscow and New Jersey. He has three American children who live with his divorced wife, and one Russian child. He considers himself "bicontinental."

The *Times* series mainly profiles people for whom mobility has conferred opportunity, and who have chosen this path and followed it well. But the demographics of migration spin a much darker tale—people who are forced to move out of desperation, or who are turned away, and whose journeys are caused by habitat degradation. There are many complex outcomes and causes. What I want to convey is the prevalence of global mobility and to consider its implications for biospheric learning.

Thomas Sowell, in *Migrations and Cultures,* places contemporary migration in a historical perspective:

In a world of 100 million immigrants—19 million of them refugees—migration is a major social phenomenon, as it has been for thousands of years. While the drama of millions of human beings migrating across the oceans of the world has been limited to the past few centuries, when modern shipbuilding and seafaring methods have made this possible, migrations of individuals and relocations of whole peoples also took place on land, and across smaller bodies of water, for many centuries before that. Thus the English of today are not indigenous to England, nor the Malays to Malaysia, nor the Turks to Turkey. Migration and conquest put them where they are.[5]

I take great pleasure in perusing the "Prehistoric Trade" map in the *Historical Atlas of Canada.* Trade goods (and hence peoples) traveled over long-distance routes that were used for thousands of years. Long before Europeans came to North America, there was surprisingly robust trading, in sync with seasonal rounds of hunting and fishing. Marine shells made it well into the heart of the continent. Obsidian from Oregon was used in British Columbia. The Adena culture of the central Ohio valley, which eventually spread up the St. Lawrence valley and into the Maritimes, was involved in a wide network of trading which included "among other items, native copper from Lake Superior, marine shell from the east coast, fire-clay tubular pipes and flaked, two-faced blades from the Ohio valley and elsewhere, and polished slate gorgets of unknown origin."[6]

The road trip is no more modern than fire. There is great ecological and cultural advantage in mobility. Rootedness and mobility represent interesting and contrasting adaptive strategies. Neither strategy is virtuous for its own sake. Both are means to attain food, shelter, and procreation.

Pleistocene wanderings may provide an evolutionary context for contemporary global migration, but the historical patterns lead to very distinct outcomes. Essentially we're working with dramatically different inhabitation ratios (people per square mile) and a whole different dimension in speed. When the Beringia land bridge was in place, it took Asians several thousand years to settle most of North and South America. You can cover that distance today in several hours.

Here are some figures to chew on. Of the estimated 100 million migrants and 20 millions refugees in the world today, 10 million people have left their homes because they can no longer make a living from their land. The International Organization for Migration estimates that 60 million people are left landless or homeless because of environmental destruction. Eighty million migrants live in cities and each year another half million move there. There are over 35 million migrant workers.[7] I don't know how you even begin to assimilate figures so staggering. Approximately one out of every sixty people living on earth is a migrant, and the overwhelming majority of these people are literally uprooted. The most obvious question to ask is, where will they go?

The stark demographics of the twenty-first century portray restless populations of rich and poor in relentless motion. Currents of people cross oceans and continents like the jet stream. During the course of your daily travels, depending on where you go and whom you're willing to talk to, you are just as likely to encounter a migrant or a refugee as you are a global trader. This is as crucial an issue for interpreting global environmental change as biodiversity or global warming. There is no room for placed-based environmental learning unless it deals with the movement of people *and* species.

Tracks around the World

Scott Weidensaul begins his fine book about migratory birds, *Living on the Wind,* by describing the summer bird residents of the Izembek National Wildlife Refuge in Alaska. Located on the shores of the Bering Straits, on the western coast of Alaska, this maritime Arctic environment with its abundant insects and fishes is a birdwatcher's paradise. It turns out that Beringia is still a crucial migratory crossroads. The prehistoric mammalian land bridge is now a cloverleaf of ornithological passage. From this place, the following migrations occur:

In Beringia, a naturalist may find Hudsonian godwits bound for Tierra del Fuego and bar-tailed godwits headed for New Zealand; the small greenish songbird known as the Arctic warbler, which migrates to the Phillipines, and Wilson's warbler, yellow with a black cap, which flies to Central America. There are fox sparrows and golden-crowned sparrows that winter in Pacific coastal woodlands, and gray-cheeked thrushes that travel to the Amazon. All morning I had been watching wheatears, black and white song birds that look like slim thrushes. They'll join their Siberian brethren and fly to China and India, then continue on to eastern Africa together, while wheatears from the eastern Arctic swing across Greenland and Iceland to reach western Africa by a European route, embracing the world in a wishbone of movement.[8]

The short-tailed shearwater has a circuitous route that amounts to nothing less than an annual tour of the circumference of the Pacific Ocean. Covering almost 35,000 kilometers, following the prevailing Pacific wind patterns, the shearwater breeds in southeast Queensland, and then flies northward along the Asian coast, past Kamchatka, across Beringia, down the western coast of North America, and finally back to Australia. Some pelagic birds cover so much oceanic territory that their breeding grounds are unknown and their populations cannot be adequately estimated.

One can hardly imagine a more transient life. The specific feats of some of these birds—the great distances they travel, the routes they take, the physiological requirements for their journeys, the navigational accomplishments—embody unusual tales of adventure. Indeed, it is a great privilege for humans to be able to trace and follow these migration stories. The evolutionary and ecological origins of these behaviors remain mysterious, although it is clear that birds will fly to where the food is. How they learn to do this, how it's genetically encoded, and the extent to which the behavior is learned, inherited, and taught are all subject to comprehensive ornithological research. You needn't travel to Beringia to observe these stories. Wherever you watch birds in flight, you are witnessing a form of ecological transience.

Bird migration is an interesting biospheric phenomenon in that it encompasses ocean currents, prevailing winds, weather systems, phytoplanktonic blooms, insect cycles—assortments of interlocking movement patterns. By watching birds, you learn to observe weather patterns. By watching what they eat, you figure out where the insects and seeds are. There are chains of ecological and climatological causation embodied in the movement of birds, insects, and seeds.

Many other mammals have interesting migration patterns. Some bat species migrate as far as 2,500 kilometers. In late April, almost in tandem with the return of the house swallow, little brown bats dart around my house at dusk, hunting, I suppose for black flies. Unfortunately, neither the bats nor birds reduce the black fly population enough to prevent them from swarming over my body as I plant the peas. In the American southwest, hundreds of thousands of free-tailed bats migrate to their wintering grounds in Mexico. You can watch them leave their caves en masse.

In his book, *Animal Migration,* Talbot Waterman reports that flyers from six different orders of insects migrate distances ranging from 1,000 to 8,000 kilometers.[9] These includes ladybugs, leafhoppers, large milkweed bugs, and desert locusts (perpetrators of awesome crop damage). In chapter 3, I described the extraordinary migration story of the monarch butterfly. The most mysterious migrants of all (because it's so difficult to trace their movements) are swimmers, including squid, shrimp, turtles, lobsters, crabs, and many fish species. The migration routes of salmon and whales are well publicized, and the subject of many environmental controversies.

The giant bluefin tuna makes almost a complete annual circuit around the North Atlantic Ocean. Their main breeding ground is the Gulf of Mexico. In May and June, they follow the Gulf Stream north to their feeding areas off the New England coast, and then in August swim further northeast across the Atlantic to the North Sea, a total one-way trip of 7,500 kilometers. In September and October they migrate to their wintering grounds off the Canary Islands and the coast of Africa. From there, it's on to the Brazilian coast, the Caribbean, and then the Gulf of Mexico. Different-sized tunas take different routes along this migrational megacircuit.

Fleeting Communities

These diverse migration stories demonstrate the ecological and evolutionary origins of migratory behavior. Some species stay put (relatively speaking) and some travel great distances. For the purposes of this chapter, what's most relevant is how the prevalence of migration influences one's conception of community. Among the most controversial concepts in the history of ecology is the notion of community—how it is defined, how it changes, whether it goes through distinct stages. In large measure, these controversies reflect the dynamics of scale.[10] What space and time boundaries are you using and why have you chosen them? Defining community in ecology, as in sociology or political science, as in everyday life, is bound to be controversial because the determination of what to include and exclude is largely interpretive. How does the prevalence of migration influence how we think about ecological community? And how does the idea of ecological community influence the meaning of being place-based?

I find Richard Huggett's comments on community in his textbook *Environmental Change* of particular interest as he examines the concept from a biogeographic perspective. He defines *community* simply as "a group of interacting individuals belonging to different species." Huggett suggests that the species in a community "occupy a mosaic of landscape patches, patches of different age supporting a different successional 'stage' of community evolution." In other words, there are so many variables in community life, species interactions, and biogeochemical impacts, that depending on what slice you are looking at, and over what time frame you do the looking, you are likely to encounter different community structures. This is a lesson in ecological transience: "Communities, like individuals, are fleeting. Species abundances and distributions constantly change, according to its own life-history characterstics, largely in response to a capriciously vicissitudinous environment . . . With each species acting individually, it follows that communities, both local ones and biomes, will come and go in answer to environmental changes."[11]

Think about it this way. Imagine a school (Antioch New England Graduate School where I teach works this way) in which different programs schedule classes on different days of the week, but also overlap. Psychology comes on Mondays and Thursdays, Environmental Studies comes on Thursday and Fridays, Education comes on Fridays and Saturdays, and so on. Depending on the day you visit the school, you're going to find some very different situations. Yet the overriding structure of the place remains the same, and events which affect the entire school also impact each program.

The hermit thrush resides in the New Hampshire woodland from May through September. The spring ephemeral wildflowers (those that bloom early, before the onset of tree cover) blossom between April and June. The migrating songbirds are just as much members of this ecological community as the year round residents, even though they spend less time there. A range of interlocking, seasonal characteristics suggest overlapping, emergent community structures. An extraordinary weather phenomenon, or a major land use decision will affect all of these communities. I've just described some of the "regular" species. What about invasive species? What is their relationship to all of this?

First, consider this caution. Fleetingness must be balanced with the remarkable constancy of some long-term community variables. Huggett notes that "there is evidence to suggest that communities stay stable for lengthy periods, say a few million years, when environmental changes are small and slow."[12] He's referring to periods when there's a relative absence of climatic fluctuation, astronomical events, opening and closing of land bridges, or tectonic activity.

However, "fast environmental changes mean hectic times in the biosphere."[13] Long-standing communities may be radically altered by landscape change or biotic invasions. "Community assembly and disassembly rates are constrained by the migration rates of component species."[14] This is precisely the contemporary situation. Radical land use changes, habitat degradation, global warming, and environmental pollution serve as catalysts for rapid environmental change. In the wake of profoundly transformed landscapes, the residents, whether year-round or migratory, must either adapt, or find some other place to live. Migration is hastened by environmental change, but not all migrants will survive.

Have a look outside your window and see how many species you can identify that have arrived by virtue of "hectic times in the biosphere." Most notable are the so-called invasive species that travel along with migrating humans—California is covered with non-native grasses, dandelions adorn American roadsides, zebra mussels clog the Great Lakes, purple loosestrife fills New England wetlands. It's speculated that the increase in mountain lion attacks in the American West is caused by encroachment on their habitat. They have more contact with humans than they are used to. In the face of ecological displacement, there is great concern about the fate of species that are unable to migrate because they lack mobility, or for migrating species that counted on diverse food sources on their itinerary, only to find them missing. New migrations may occur bringing species to habitats where they have no historical and ecological precedent. What environmental changes will follow in their wake?

This survey of migration and community, in relationship to environmental change, is intended to offer an ecological perspective on the movement of peoples and species. Communities always consist of species that are just passing through. What matters is the rate at which this movement occurs and how it affects the long-term residents. The twentieth century saw an unprecedented increase in the movement of human populations. And with the accompanying impact on the ecological landscape, it is clear that human migration must be viewed in broader ecological terms. For what awaits us in the twenty-first century is the challenge of widespread cultural *and* ecological diasporas. Those of us who are constantly on the move have much to learn from the conditions of a diasporic residency.

A Diasporic Residency

The widespread, global migrations of peoples and species is the shadow of globalization, returning to haunt the sheltered domicile of seemingly affluent domains, knocking on the doors of walled communities, a grim reminder of the transience engendered by forced uprootedness. As migration accompanies globalization, the displacement of peoples and species becomes more pronounced.

Historically, when migrating groups of people maintain the integrity of their culture despite their dislocation, we describe a *diaspora,* defined by Chaliand and Rageau in *The Penguin Atlas of Diasporas* as "the collective forced dispersion of a religious and/or ethnic group, precipitated by a disaster, often of a political nature." These groups maintain their cultural integrity via collective memory, which "transmits both the historical facts that precipitated the dispersion and a cultural heritage."[15]

How do traditional place-based communities retain their ecological knowledge when they no longer have access to their homeland? In the twenty-first century we face the prospect of multiple ecological and cultural diasporas, millions of migrants attempting to salvage their ecological and cultural integrity, using whatever means necessary to maintain some form of collective memory, or confront the dire consequences of naked assimilation and the loss of cultural identity.

Anthropologist James Clifford writes that "the language of diaspora is increasingly invoked by displaced peoples who feel (maintain, revere, invent) a connection with a prior home."[16] A diasporic tradition lends geographic and historical coherence to the condition of uprootedness. Even though you are a temporary resident in your new place, by tracing your path and linking it to myth, political circumstance, homeland, and culture, you have a means to contemplate both roots and destiny. Hence, Clifford observes that "many groups that have not previously identified in this way are now reclaiming diasporic origins and affiliations."[17]

This diasporic idea encompasses a wide range of peoples and situations, including political and religious persecution, economic opportunism, and urban cosmopolitanism. And it includes historical diasporas that entail several thousand years of movement (the Jewish people) and others that are more recent (Vietnamese boat people). Given this range of circumstances, what boundaries can be placed around the diaspora concept and why is it relevant for place-based learning? Robin Cohen, a sociologist, and an eminent migration/diaspora scholar, summarizes nine characteristics of diasporas:

(1) dispersal from an original homeland, often traumatically; (2) alternatively, the expansion from a homeland in search of work, in pursuit of trade or to further colonial ambitions; (3) a collective memory and myth about the homeland; (4) an idealization of the supposed ancestral home; (5) a return movement; (6) a strong ethnic group consciousness sustained over a long time; (7) a troubled relationship with host societies; (8) a sense of solidarity with co-ethnic members in other countries; and (9) the possibility of a distinctive creative, enriching life in tolerant host countries.[18]

These diverse criteria reflect the complex historical and cultural dimensions of the diasporic phenomenon. Moreover, there may be crucial class differences among diasporic populations. Clifford reminds us that in distinguishing between "affluent Asian business families living in North America from creative writers, academic theorists, and destitute 'boat people' or Khmers fleeing genocide, one sees clearly that diasporic alienation, the mix of coercion and freedom in cultural (dis)idenitifcations, and the pain of loss and displacement are highly relative."[19] So a diasporic tradition may entail feelings of "loss, marginality, and exile," while it coexists with "adaptive distinction, discrepant cosmopolitanism, and stubborn visions of renewal."[20] Interestingly (please recall the material on hope and foreboding in chapter 3) Clifford claims that "diasporic consciousness lives loss and hope as a defining tension."[21]

By identifying with a dispersion movement of peoples, which may include a global diffusion pattern, a diasporic awareness involves tracing a global path of movement. For diasporic peoples, a place-based orientation necessarily includes mobility and uprootedness. They live in between places, or reside conceptually in several places at once—a symbolic homeland and an adopted host.

A diasporic residency lies on the margins of local and global. It expands the idea of homeland and it calls into question the meaning of native. Who belongs in a place and what criteria define the terms of belonging? Ownership of property? The longevity of your residence? Your economic contribution? How much food your garden produces? Your influence on local politics?

In a world in which uprootedness is such a crucial demographic factor, what claim can anyone make to being native? I like to think that my twenty years of residency on these five acres of New Hampshire woodland lends me a special local claim and that I've attained this through proximity, familiarity, intimacy, and attitude. Sometimes I feel more rooted than my newest neighbors, as if my lengthier stay and my ecological pedigree grant me superior title. But how does the old-timer down the road view me? I will forever be a newcomer, and so will my children, and probably their children, too. And surely that old-timer can't be considered indigenous to this place. Who knows what combination of usurpation and chance granted him the privilege of a longer-term residency?

Who are my people and where is my homeland? Are they my neighbors in this Monadnock woodland? My relatives who are scattered around North America? Other Jews (including relatives) who live in Israel? In 1990, I visited Moscow to set up environmental exchange programs. The day I was leaving, there were thousands of people in the airport getting ready to emigrate. I was told they were Jews fleeing to Israel because they were concerned how Russia's instability might impact them. All of my grandparents were born in Russia. How little separated me from these emigrants—merely the vagaries of fate and history. In some ways I had much more in common with them than I do with my New Hampshire neighbors, yet we are worlds apart.

I have no desire to return to Russia or to live in Israel. I feel there are lifetimes of exploration within a twenty-mile radius of my home place. But I should be very careful about asserting false claims of residency. I have a very tenuous hold here and so do my neighbors. Just like the juncos who made a brief appearance by my feeder yesterday on their migration north, I am just passing through this landscape. The junco is searching for food. I'm searching for history. No matter how much this landscape molds and shapes me, it can only modify my diasporic origins. I have an adopted homeland, and my people have a history of transience.[22] But the more carefully I look, I begin to realize that the same is true of many of the people and species with whom I share this place. It depends on the scale of one's view. We are all relative newcomers, and perhaps we should be sufficiently humbled together.

Yet I am not willing to let go of this place-based philosophy. Not only does it make good educational sense but it speaks to the possibility of ecological fidelity, and lends me a sense of rootedness (however transient) in a world of ceaseless motion. A diasporic residency conveys a special responsibility. It requires me to understand the confluence of habitat and history. I can only come to know the special qualities of this place by studying the details of its *natural* history, what species came before me, which live here now, and which are just passing through. But this knowledge has a context. How did I come to be here? What people and species do I share this place with and what are the ecological terms of our common residency? What traditions do we share in common? What is it about this place and our connections to it that endures?

The Confluence of Habitat and History

Just as transience portrays one dimension of human and ecological history, permanence reflects another. These concepts are best understood in reference to each other. Humans build monuments as trees require roots. Homelands sanctify lineage. Permanence is important because it implies stability, belonging, and continuity. It contributes to the perception of something lasting indefinitely, providing a basis for moral and ethical conduct that supersedes the limited context of your time. So permanence is embodied in different ways—ancestry, landscape, language, deity—yielding concepts, institutions, and belief systems that intend to endure. Permanence is a search for faith and order, embedded in landscapes and communities that undergo continuous change.

For the diasporic people whose place is dispersed on the winds of history, where might rootedness reside? In Judaism, permanence resides in the Torah, what has been described as a "portable homeland," a legalistic and spiritual place that has profound meaning wherever the book is studied. On Shabbat (the Sabbath), Jews throughout the world know that wherever they might be they have the weekly reading of the Torah to guide them. Its truths and liturgy transcend the limits of physical geography, placing the Jewish people in the context of time rather than space. The Torah endures from the dawn of human civilization.

Abraham Joshua Heschel explains how "Judaism teaches us to be attached to holiness in time, to be attached to sacred events, to learn how to consecrate sanctuaries that emerge from the magnificent stream of a year."[23] The Sabbath represents the ritualistic sanctification of time, recurring on a weekly cycle, enduring across space in multiple cultural settings. "Six days a week we live under the tyranny of things in space; on the Sabbath we try to become attuned to *holiness in time.* It is a day on

which we are called to share in what is eternal in time, to turn from the results of creation to the mystery of creation; from the world of creation to
the creation of the world."[24]

Yet time represents only half of the picture, for we neglect space (and
place) at our ecological peril. If Heschel's plea for mindfulness in time is
to be heeded and sustained, it must not be balanced on "the tyranny of
things in space." Material life is not merely the exploitation of place for
the purposes of making a living. It is also a means for contemplating creation. Heschel is ambivalent on this point, at times intimating that space
entails the enslavement to things, but also envisioning both space and
time as the works of creation. Humans dwell in a place (in space) and
time. Whether your history entails diaspora or rootedness, the search for
permanence necessitates the sanctification of both place and time.

Place is neglected at the expense of habitat. What happens to place
when there is no longer any ecological endurance or continuity, when
the coevolution of people and landscape is sharply severed, when the
stories of inhabitation are no longer told, relegated to a library as primi-
tive idiosyncrasies? The stories and traditions of a place—its mysteries
and secrets—are recorded in the cultural artifacts, the living language,
and the subsistence practices of people who have lived there for multiple
generations.

David Abram persuasively suggests that the interpenetration of
species, people, and landscape is the basis of local language. "If we listen, first, to the sounds of an oral language—to the rhythms, tones, and
inflections that play through the speech of an oral culture—we will
likely find that these elements are attuned, in multiple and subtle ways,
to the contour and scale of the local landscape, to the depth of its valleys
or the open stretch of its distances, to the visual rhythms of the local
topography. But the human speaking is necessarily tuned, as well, to the
various non-human calls and cries that animate the local terrain. Such attunement is simply imperative for any culture still dependent upon for-
aging for its subsistence."[25]

Stories of inhabitation emerge from a profound familiarity with the
landscape and all of its residents. Only the proximity that comes with
generations of knowledge can begin to reveal their depth. A place is like
a fractal. It recursively unfolds, always revealing insights and wisdom.

Time is neglected at the expense of history. What happens to time
when the history of place is reduced to a human story, the timetables of
lineage and power, an angle for the evening news consolidated into the
span of a sound bite, and the evolutionary expanse of Earth's swirling

history is reduced to a National Geographic video? To understand the origins and transformations of a place, one must always travel into deep time. Such a journey requires the scientific stories of biospheric evolution, indigenous creation myths, and perspectives of the world's great wisdom traditions. "Very deep is the well of the past. Should we not call it bottomless?" (Thomas Mann)[26]

As a place-based environmental educator, reared in the throes of Diaspora, I find that I am drawn to two types of stories, those of the landscape where I now reside, and those of my diasporic past. It's the stories of this bioregion that yield the tradition I crave. I study the acorns to learn the story of oak. I follow the watershed to learn the story of the hydrological cycle. Observing winter tracks and animal behavior helps me understand what survival means in northern New England. Meandering along stone walls helps me decode the human ecological history. And when the hermit thrush returns for the summer and sings its indescribably beautiful song, I listen carefully for the story of its migration. I aspire to study the language of the birds. These stories inform the present moment and sanctify this place in a timeless way. Might a sunrise, the first blooming flower of spring, the peepers' calls, the falling of snow, the turning of leaves, the departure and arrival of songbirds, also be perceived as the holiness of time, sacred events in their way?

In *The Cultures of Habitat,* Gary Nabhan remarks that tracking down the stories of plants and animals is crucial to inhabitation, and a means to inform and enrich how we learn about a place.

Try to imagine the still-untold stories, the sudden flowerings, the cataclysmic extinctions, the episodic turnovers in dominance, the failed attempts at mutualistic relationships, and the climaxes that took hundreds of years to achieve. In every biotic community, there are story lines that fiction writers would give their eyeteeth for. Desert tortoises with allegiances to place that have lasted upward of 40,000 years, dwarfing any dynasty in China. Fidelities between hummingbird and montane penstemon that make the fidelities in Wendell Berry's Port William, Kentucky seem like puppy love. Dormancies of lotus seeds that outdistance Rip Van Winkle's longest nap. Promiscuities among neighboring oak trees that would make even Nabokov and his Lolita blush. Or all-female lizard species with reproductive habits more radical than anything in lesbian literature.[27]

Right next to these stories, these place-based learning experiences, I also crave the stories of the Jewish Diaspora. They too provide moral and ethical guidance, teaching me about the history of the human community with all of its sufferings and satisfactions. Here are stories of slavery

and liberation, freedom and exile, materialism and simplicity, friendship and betrayal, conviviality and antagonism. Behind the specific inci-
dents and circumstances of these diverse historical settings, there are
prayers and practices that endure, providing me with solace and faith,
nourishing my soul as deeply as the first spring song of the hermit
thrush.

Some stories are snippets, snapshots of indigenous people, species,
and landscapes, read as short poems that leave lasting impressions. Others follow you around. They move from place to place, modified by vernacular settings, interpreted on a universal stage. The diasporic is a
messenger, carrying stories of many places, linking the local to the global
by virtue of her wandering.

Place-based transience can be explored at the confluence of habitat
and history—how the story of life passes through this moment in place
and time. The permanence of place is forever elusive. Perhaps it can only
be grasped in the ephemerality of the present moment, when the full
mystery and splendor of creation is fleetingly revealed.

Place-Based Transience

In the winter of 1998, during the darkness of the winter solstice, a gentle
but persistent icy precipitation fell continuously for four days and
nights on the hilly terrain of southwest New Hampshire. Sometimes
warm and cold fronts get tangled in an endless tug of air masses and they
just stay in place for a while. This time they remained stationary over
central New England, yielding fog and ice, and now-legendary power
blackouts.

With a good woodstove your inconvenience is sufficiently reduced
and once you work through the uncommon disruptions of schedule and
routine you realize that the absence of electricity has some distinct ad-
vantages. Far more unnerving is the incessant crackle and crashing of
falling branches in the woods. As I lay in bed, my ears tune to the acoustic
dimension of this storm. Every minute or so, with each soft breeze, the
percussive pattern resumes—tinkles, clicks, snaps, whooshes, thuds,
and crashes—high hats and snares are the acoustic cymbals of this extraordinary event. The forest undergoes a purge and it matters little
whether power lines, homes, or cars are in the way. As each tree falls I
consider the utter tenuousness of my place and how a natural catastrophe could easily dismiss what I worked so hard to build. Yet given the
appropriate time scale, events like these are merely part of a natural cy-

cle. This is the type of damage from which the northern forest will eventually recover.

On a Saturday morning the weather finally breaks. The day is mild, sunny, and a very deep blue. I take a long walk to survey the ravaged landscape. Power lines are down everywhere. Large branches are perched precariously in contorted configurations. I need a helmet to walk safely as ice and branches continue to plummet. As the sun reaches its midday height the forest glistens as I have never seen it, a dazzling spectacle of illumination from the ice-covered trees and shrubs—mirrors and jewels endlessly reflecting as sparkles of frosty glitter. What a strange juxtaposition. Forest fragments lay strewn in the midst of an unprecedented display of glamour. I know I might never again see such transient conditions. The impermanence of biospheric creation is revealed in the midst of this short-term meteorological event, so fraught with devastation.

Never had my own transience appeared so tangible. In such moments I am grateful for the opportunity to observe and experience such magnificence. As an ecological transient I am just passing through both this landscape and this life, on the foundation of grand biogeochemical cycles, a gift from earth's planetary physiology. I am a participant in and observer of the subtle and dramatic transformations of life and landscape which allow me to reflect on (im)permanence in the Buddhist sense, just another manifestation of conscious awareness. And Heschel would say that this expression of the ineffable is a pathway to creation. I need this Jewish-Buddhist-ecological view as a way to endure in the face of events that demand some kind of interpretation and meaning.

Ironically, the ice storm was discussed locally in the context of the unusual winters we've been having. What greater threat to permanence can there be than the prospect of global weather change? You expect the weather to conform to certain patterns and when those patterns stop making sense you are apprehensive about the consequences. Consider the ice storm in the broader context of habitat transformations that are sweeping the globe—the unprecedented fires in the tropical rainforests, the clear-cutting of boreal forests.

These are the circumstances that impel place-based ecological learning—the intention to protect the last stand, to make sure that the integrity of a habitat is sustained, to allow the ecosystem to endure, to promote cultural and biological diversity. How can this be achieved under the aegis of global economic expansion, when habitats and cultures are fragmented at such a relentless pace, when people and species must abandon their homelands, when place attachment is so difficult to develop?

Good environmental education provides a spatial and temporal context for ecological learning. As you change perspectives and scales you recognize the shifting parameters of cultural and biological diversity. There is always more to a place than meets the eye. Depending on the scale of your observation, different species move through a place, or reside there, in various ecological and historical circumstances. The life of a community evolves through its landscape.

As a Diaspora Jew, I must fully admit to the circumstances of my residency. Despite my best intentions to remain here and my deepest hopes that my children settle here, and their children too, I have no idea how "permanent" my family's residency may be. I can never be fully grounded here. I must accept that. But perhaps I can come to appreciate what this place provides. This entails learning as much as we can about our place, opening our hearts to the lessons of the landscape, listening to the stories of the land and its inhabitants, and understanding the circumstances of their arrivals and departures. It is in the comings and goings of people, landscape, flora, and fauna that this place is connected to a broader and deeper ecological experience. One's transience merely puts place in perspective—it enriches the process of passing through. While we are here, whether for a week or a dozen generations, it is our responsibility to help this place endure. And perhaps, as a diasporic people, we can contribute the wisdom of our traditions to the mix. Then we might begin to belong.

A cosmopolitan diasporic learns how to stand in several places at once, developing multiple fidelities and allegiances. You can be place-based locally while cultivating deep concern for a broader spectrum of global movement. What's crucial is the depth of the connection. As I come to know the phoebe who resides in my garage I must care about all of its habitats and feeding grounds, even though I will be most responsible for its proximate residency. And on Passover, when I contemplate the fate of Jews in Exodus, my concern must not only be for displaced Jews and our common history but for displaced peoples who happen to be my neighbors. The diasporic becomes rooted by virtue of connecting the relationships between people, species, and places.

A Wild Diasporic Garden

The etymology of *diaspora* is particularly revealing. It means "to sow, scatter" (from the Greek), the movement of seeds. In ecological terms, could we say seed dispersal? Fungi, plants, and mammals have a complicated and elegant evolutionary history of entangled forms of seed dispersal. Note Lynn Margulis and Dorion Sagan's comments about fruit. "Like colorful grocery packaging, bright and flowerful fruits with inedible or discardible cores manipulate the animal into collecting and spreading the offspring of the plant."[28]

Dandelion spores ride the wind and settle America's lawns and roadsides. Birds carry various seeds and spores miles aloft, spreading them great distances from their original habitats. Weidensaul reports that in Central America, the fruits of several species of trees ripen, neatly timed to the waves of northbound migrants.[29] My grandparents rode the ocean waves from Europe to America, sowing their seed as well, scattering their people and heritage.

Seen this way, the world resembles a wild diasporic garden—the inevitable, irrepressible intermingling of people, species, and landscapes. To say "wild" lends diaspora a vitality and essence both mysterious and unknowable, beyond human control and cultivation. To say "garden" implies an intentional, organizational plan. Does wildness imply transience and the garden permanence, a way of placing temporary order on nature? Observing and studying wild migrations, whether they be tuna, bats, warblers, or humans, is a reminder of the unbounded movement of the biosphere. Calling particular human migrations "diasporas" lends cultural coherence and a semblance of order to a condition of uprootedness. The term "wild diasporic garden" implies a continuous tension between rootedness and mobility, wild and settled, local and global, transient and permanent.

I have borrowed the garden metaphor from Robin Cohen, who develops a tongue-in-cheek taxonomy of diasporas, linking them to gardening terms. What's tongue-in-cheek to a sociologist may be taken too seriously by an environmental educator. But let's explore this concept further to see what it yields. As a migration scholar, Cohen is interested in the special qualities of diasporas as forms of long-term migration. He intends to interpret the dimensions and implications of diasporic identity. Cohen and Clifford (see above) are particularly interested in this diasporic identity for two reasons. First, it reflects the demographics of globalization, thus providing insight into cultural coherence in turbulent times. Second, it may provide a better understanding of how to link local and global identity, perhaps leading to new forms of global citizenship. Cohen observes that diasporas "bridge the gap between the individual and society, between the local and the global," hence performing a "vital social role." "The sense of uprootedness, of disconnection, of loss and estrangement, which hitherto was morally appropriated by the traditionally recognized diasporas, may now signify something more general about the human condition. Why not celebrate the creative, enriching side of living in 'Babylon,' the radiance of difference?"[30]

Can this diasporic awareness also provide a means to perceive the necessity of cultural and biological diversity? This is the question of most direct relevance to environmental education, and is the conceptual link between migration studies and place-based environmental learning. Let's look further at the wild diasporic garden.

Cohen's "good gardener's guide to diasporas" (sounds like an intriguing seed catalog!) interprets the history and sociology of diasporas by using gardening terms.[31] Weeding corresponds to victim/refugee diasporas, the subjective notion that there are too many individuals of a species. Just as a gardener may uproot, cast out, or dispel the weeds using "weed killers," a nation may resort to expulsion, deportation, or genocide (ethnic cleansing). Interestingly, what a gardener perceives as weeds may serve a vital function to wildlife and be crucial to the ecology of a region. A most recent case in point is the eradication of milkweed by American farmers. Milkweed serves a vital pollination role for monarch butterflies (see chapter 3), and poses very little threat to cropland. The cultural and ecological interpretation of what constitutes a weed is of crucial importance in understanding biodiversity.

Sowing refers to imperial or colonial diasporas, opportunistic movements to gain economic and political holds in new territories. Here one interprets seed dispersal literally. You scatter seed to place roots in diverse locations. In this case, depending on the relative strength and weakness of the seeds in relationship to their new environment, the new host may experience the dislocation. Diasporas move both ways, impacting both the migrant and the more permanent resident. In ecological communities, this process emerges in plant succession dynamics, where particular species are adept at settling new areas, sometimes leaving a lasting imprint, sometimes one more ephemeral, but in many cases, preparing the landscape for additional waves of dispersal and settlement. The historical geographer, D. W. Meinig uses a variation of this metaphor to describe the European colonial settlement of North America.[32]

Cohen links transplanting to labor and service diasporas, in which migratory movement consists of people who move from one place to another primarily in search of work. He is referring mainly to low-income workers, indentured servants, and unfree labor. These forms of cheap labor, typically supported by wealthy entrepreneurs, accounted for a great deal of the international migration of the nineteenth and twentieth centuries, typically causing social and political turmoil, especially between workers in the host country and the migrants. Much of the bias against immigration is legitimately tied to these forms of imported labor, although the migrants themselves are victims of circumstance and inappropriate targets. Cohen refers to this as "digging up and replanting." The "layering" diasporas represent the trade-business-professional movements, which Cohen likens to "establishing trade outposts or branch plants," a common strategy of merchants and entrepreneurs. Layering suggests "taking cuttings without separating them from the parent plant until they are rooted."

Cross-pollinating refers to cultural and hybrid diasporas, in which the mixing and mingling of peoples is most pronounced. In these cases, we are likely to observe an intrinsic cosmopolitan quality to migration, with a finely integrated mesh of host and migrant. Cohen reminds us that "fruits cannot develop from flowers without adequate pollination" and that "better crops arise from additional pollinator varieties." Just as water and wind carry plant pollen, so do cultural diasporas ride the ocean waves (physical movement) and the air waves (the movement of ideas or music).

This metaphor and typology is mainly useful as a means of imaginative expression, not as a serious analytical tool. Here is another learning path for perceiving global environmental change—how the movement of people and species connects your local place to global migratory processes. As global residents in a wild diasporic garden we may wish to consider which gardening techniques best serve ecological citizenship, which provide a way to understand uprootedness in relationship to species extinction, global warming, and habitat degradation.

If you perceive the biosphere as a wild diasporic garden, which metaphors of diaspora might form the basis of global environmental policy, ecological restoration, and immigration policy? Such a complex question is way beyond the scope of this book, yet crucial to local decision making and among the most controversial of environmental issues (Sierra Club immigration policy). The educational implications of this question can surely be applied to place-based environmental learning. How might you tend your local community garden (whether it's carrots, purple loosestrife, or Cambodian refugees) given the mix of residents and diasporic people? Perhaps we need more pollinators and fewer transplants, or we should reconsider the value of weeds. What lessons can be derived from closely observing ecological migrations and how might they inform our perception of transients?

Learning to perceive global environmental change entails detailed observations of what moves through your place and noting the terms and conditions of their arrivals and departures at all points along the way. Which migrations are customary and which are extraordinary? Which result from unusual ecological events and what might their impact be on your local place? What people and species will pollinate your garden and which threaten to leave it plundered and pillaged with you being displaced in turn?

Robert Kaplan, in *An Empire Wilderness,* a study of the North American Southwest, shows how you can no more build walls to separate borders than you can erect enormous nets in the air to halt the migration of birds. The influence of Mexicans and Americans on both sides of the border is too powerful a force to be halted by immigration laws.[33] Stopping the flow of people is like plugging a dike. People and species always find a way to get through. The wildness of migration will transcend its taming as ecological and economic conditions dictate the terms of the journey. What's most important is how communities can best prepare for both migrants and longer-term residents alike so there's a balance between who comes and who goes. This will only occur when local places understand that the integrity of their place is never assured unless there is compassion and concern for all places on the migration trail.

There are many interesting environmental exchange programs that pave the way for this kind of learning. In chapter 6, I described several bird and butterfly programs that link correspondents along all paths of a migration route. Weidensaul tells the story of Hawk Watch, a local Pennsylvanian raptor ecology program that extended its reach to Veracruz, Mexico, the most extraordinary raptor flyway in the world, where on a given autumn day, over a million hawks might pass through. He explains how an educational exchange program has enabled citizens in both the American Northeast and southern Mexico, to take a collaborative pride in the sanctity of this migration, linking local conservation practices to this international affiliation.[34] Shouldn't we aspire to afford the same courtesy, respect, and cooperation to displaced peoples as we do migratory hawks? Both are crucial measures of a place's commitment to promoting ecological community.

Every community is a garden in the wild diasporas of migration, containing "traveling cultures" of migrating birds, airborne dandelion seeds, or immigrants in search of a better life. We choose how to cultivate that garden by virtue of our relationship to the land, the species that dwell on it, and those that are just passing through. In a wild diasporic garden, there is a correspondence between biological diversity and cultural diversity, and pollination is the means to maintain the fertility of such ecological ground. Let me place my roots in a cosmopolitan community of species, in the confluence of habitat and history, with a fertile mixture of wisdom traditions, compelling ideas, endemic species, among a global network of seeds and spores.

A Cosmopolitan Community of Species

In his groundbreaking essay, "Missing the Boat: Why Cultural Diversity Didn't Make it onto the Ark," Gary Paul Nabhan makes the important observation that so-called biodiversity hot spots occur in areas where cultural diversity also persists. He shows that there is a direct correspondence between the prevalence of biodiversity, the long-term coexistence of diverse cultures, and the presence of multiple languages. "Of the nine countries in which 60 per cent of the world's remaining 6,500 languages are spoken, six of them are also centers of megadiversity for flora and fauna: Mexico, Brazil, Indonesia, Zaire, and Australia. Geographer David Harmon has made lists of twenty-five countries harboring the greatest number of endemic wildlife species within their boundaries and of the twenty-five countries where the greatest number of endemic languages are spoken. These two lists have sixteen countries in common. It is fair to say that wherever many cultures have coexisted within the same region, biodiversity has also survived."[35] Nabhan conversely suggests that "wherever empires have spread to suppress other cultures' languages and land-tenure traditions, the loss of biodiversity has been dramatic."[36]

From an entirely different quarter, Steven Pinker, a linguist, describes the global extinction of indigenous languages. Approximately 3,600 to 5,400 languages may well be lost in the next century, what amounts to 90 percent of the world's total.[37] What is the significance of this for biodiversity studies? Despite the best attempts of conservation biologists and ecologists to collect natural history data, the most important *knowledge* regarding natural history is locked in the languages of indigenous peoples who study proximate species over dozens of generations. As the language is lost, so is the tradition of direct, visceral ecological and evolutionary learning. Languages are a species of sorts and their extinction is inextricably linked to the loss of biodiversity.

Conceptual links such as these cross many interesting ethnobotanical, anthropological, and ecological frontiers. Perhaps they presume too much about the great diversity of Pleistocene experience and represent as much projection as analysis. Yet I find great elegance and virtue in this interesting correspondence—cultural and ecological diversity is a requirement of biospheric integrity, human vitality, and enduring inhabitation. In these dense zones of species mixing you imagine fecundity and cross-pollination, derived from endemic knowledge and migratory flows, the coevolution of rootedness and movement. An intricate versatility emerges—improvisations on ecological and evolutionary experience—using the scale of landscape. There is no greater testimony to biospheric creation and the human capacity to observe it than the celebration of diversity.

But such coexistence is never easy. It's tempting, but superficial to romanticize the bucolic intermingling of diverse peoples and species. Elaborate ecological and cultural rituals are required to determine territorial realms, resource rights, and legitimate spaces of habitation. In wild nature much of this is worked out through generations of evolutionary possibilities, the inevitable dance of cooperation and competition. Commons scholars such as Elinor Ostrom show that such arrangements emerge within long-standing human communities as well.[38] However, for humans, the challenge of heterogeneity is as old as culture—learning how to live together. It should be no surprise that the most painful and complicated controversies in American public life concern issues of race and nature. People are ambivalent about difference, often fearful of the Other, and will often go to great lengths to insure the homogeneity of their experience.

It is now a cliche to assert that a shrinking globe necessitates increased contact among peoples. All of the migration statistics I cited earlier in this chapter indicate that the movement of peoples is a vital twenty-first century demographic. But such movement doesn't necessarily entail mixing, or the emergence of a cosmopolitan mentality, a person who flourishes on the vitality of difference. One can point to all of the insipid signs of homogeneity—monocultures of crops and commodities, perpetrated by misguided notions of world order. Most experiences of globality are a complex mix of the imperial and the indigenous, the migratory and the rooted, the parochial and the cosmopolitan. There are as many examples of ethnic chauvinism as there are of global citizenship.

The "sophisticated" cosmopolitan is the person who has learned to live with difference, who thrives on the mix of ideas, cultures, and peoples who mingle in the ports, plazas, and cafés of human traffic. Aware of multiple points of view, exposed to a diversity of styles, the cosmopolitan derives insight and wisdom from these compelling influences. To be truly cosmopolitan, one must be equally exposed to and influenced by the biological diversity of the planet, the extraordinary melange of species and habitats. Cultural diversity emerges from biological diversity. That is the ultimate origin of our creative ideas—our multiplicity of mixtures, hybrids, and their countless variations. Without biodiversity, the world is merely a human reflection. Without creation, there is no creativity.

One of the great gifts of living in this time in human history is enjoying exposure to so many dimensions of human culture. Walk into a bookstore or library, cruise the Internet, and learn about any style of music, any world literary tradition, or dozens of spiritual paths. Plumb them as you will. Pick, choose, and mix accordingly. This delightful intellectual pollination is both enthralling and overwhelming.

How ironic that among the most disturbing aspects of the biodiversity crisis is the threat to global ecological pollination processes, the very intermingling of species that is the basis of human survival. Pollination is the best ecological example and metaphor we have for a cosmopolitan community of species.[39] Migrators pollinate! It is the most intriguing dimension of Robin Cohen's good gardener's guide to diasporas. One could find no more virtuous cosmopolitan statement than to restore the vitality of a community's pollinating species. Observing biodiversity is the best means we have of understanding how different species learn to live together.

Convergence

On a misty, gray May morning, I leave my desk for a while to walk the hills and survey spring. Today the frontal systems are aligned in a pattern similar to the ice storm. High pressure is stationary over the Maritimes and low pressure is stationary in the Ohio Valley. The result is a stagnant weather system, with periods of showers and fog bringing moisture from the South. But it's much warmer so there is no ice. As I walk on the dirt roads near my house, the remnants of last winter's ice storm are evident. Treetop branches are stacked neatly on the side of the road, making a fine habitat for small mammals and birds. Eventually they will become kindling. The glittery, white, ice-coated forest of just a few months ago is now sprouting lime-green shoots and leaves. The forest shimmers just the same, but with the new life of spring. The misty, damp weather triggers a surreal memory, my recurring dislocation dream. I perceive my life as a string of landscapes.

As I walk down the hill, back toward the house, I listen to the newly arrived migrating birds and how their songs blend so beautifully with the
permanent residents. A chickadee, a tufted titmouse, a downy wood-
pecker—these songs can be heard throughout the year. But in the distance, coming from the patch of trees just in front of my house, I hear the
pure, melodious flutelike song of the hermit thrush singing the news of
the landscape. And sprouting from the ground, as if by magic, surge
Canada mayflowers, painted trilliums, Indian cucumber-roots, clintonia, the enduring signs of life after another cold, dark winter. For me,
these species are the landscape of home.

On returning to my study, I glance at the bookshelves, taking solace in
the coexistence of field guides, nature writing, books about the biosphere, Jewish and Buddhist texts. These volumes help me integrate my
life on this landscape, placing in perspective my fleeting appearance
here, sustaining my gratitude for this place, helping me cope with the
sorrows of extinction, the suffering of life forms, the impermanence of
my evolving worldview. These bookshelves, this community, Mount
Monadnock, the hermit thrush, the Jewish diaspora, my family—this is
my ephemeral permanence, my portable homeland.

There is a convergence between the ecology of this landscape and the
historical patterns of diaspora. Both processes can serve as a source of
stability, as a way to recognize impermanence, grounded in the cycles of
ecology and history. I repeat the etymological mantra of transience—
just passing through. With this admission I can aspire to cultivate an
ethics of transience as criteria for permanence. What are the conditions
of habitation given my diasporic ancestry?

I know my recurring dream will return one evening. As I resume my
wandering in the streets of a vaguely familiar city, I can approach my
journey with less trepidation and face my fears head-on. There is continuity in my dislocation. I am connected to a collective anxiety dream,
stretching across landscapes and generations—the shifting tides of
place and culture. It is the matrix of these ecosystems and the cultures
they contain that sustain biodiversity. To be place-based is to affirm
membership in a cosmopolitan landscape. This membership is a source
of endurance wherever I may reside, in the forest or the city, awake
or in a dream. Now I know that my wanderings serve a purpose. My
rootedness lies in the interplay of transience and permanence forever
unfolding.

8   A Biospheric Curriculum

On May 1, in the Monadnock region of southwest New Hampshire, at an elevation of 1,200 feet, in the beech-oak-maple-birch forest, spring is poised like a coil. There are still no leaves on the trees but the red maple blossoms, the brown beech buds, and the yellow-green birch flowers add delicate pastels to the awakening landscape. The forest floor is covered with the sprouting leaves of Canada mayflowers, although their flowers and those of the pink ladyslippers, the trillium, the Solomon's seal, and clintonia—and all of the vernal photosynthetics—won't bloom for another week or two. The first migrating birds are moving through. Just a few moments ago I had a terrific view of a male yellow-rumped warbler, a flash of color from the subtropics. The black flies are already here but they aren't really biting yet. One more really warm day and they'll be a full-fledged nuisance.

We're beginning another spring cycle of biomass explosion. An orgy of photosynthesis will bring green depths and layers back to New Hampshire—a dense forest canopy, thick and fertile fields of grass, carpets of moss. Just a few weeks from now the landscape will be completely transformed. This time of year, if you are patient enough, you can literally watch the grass grow. Under the influence of such prolific growth, it is easy to be inspired by the renewal and resilience of the biosphere. Surely spring is a time to celebrate the unbounded exuberance of life. On such a fine spring day you want to share your appreciation of biospheric cycles, natural history, black flies, and spring wildflowers, and you devote your work to doing so. You have a deep faith that with just the right prompting and support such appreciation can be nourished and nurtured.

Yet why is it that some people can observe the natural history of spring for hours on end while others would be instantly bored and impatient, finding the prospect of sitting in silence intolerable, feeling naked without a Walkman, a laptop, or a cell phone? Many times I've experienced both states of mind and I know that it takes a special effort to resist the constant chatter of human artifacts. There are times when you need to be shown that the silent observation of nature has great virtue. How is this quality instilled in a person?

Howard Gardner, the developmental psychologist, suggests that there is a naturalist intelligence and some people are predisposed to be excellent naturalists.[1] From an evolutionary perspective wouldn't it make sense that our Pleistocene minds are inclined to study nature? Wouldn't that be a prerequisite of survival? Edward Wilson, Stephen Kellert, and others suggest a biophilia hypothesis, or the "innately emotional affiliation of human beings to other living organisms."[2] Have we become so "civilized" that such an obvious premise must now be hypothesized?

The things that you're interested in are molded in part by your genetic predispositions and your Pleistocene heritage, but they're also determined, and certainly catalyzed by what you're exposed to during the course of a lifetime. Taking an interest in natural history and the biosphere requires that this Pleistocene heritage is sufficiently awakened, because no matter how much people may be predisposed to thinking about nature or affiliating with life, a civilized culture may get too easily stuck in its own trappings, forgetting its origins in wildness, and its reliance on the ecosystem.

The naturalist intelligence and the biophilia hypothesis suggest that studying natural history and the need to affiliate with life are genetically based. This may be so, but it takes an appropriate educational environment to resurrect this fundamental learning capacity. There is no magic formula for cultivating biospheric awareness. Conservation biology, natural history, and biosphere studies must become priorities for K–12 and undergraduate curricula. Public awareness of global warming will remain superficial until an interdisciplinary earth systems science plays a prominent role in schooling. The biosphere will forever be an esoteric concept unless it receives the scientific, spiritual, and artistic attention it deserves.

In *Bringing the Biosphere Home,* I've advocated the importance of observing local natural history as a means of interpreting the biosphere. This is the foundation from which broader and bolder conceptual leaps might be attained. With balanced doses of analysis, imagination, and compassion, one learns how to traverse different scales of perception.

I've suggested that biospheric perception is both an ecological and existential challenge, revealing depths of learning that are both wonderful and unfathomable, demanding intellectual and visceral journeys that move from one's backyard to the history of life on earth. This is a perceptual challenge that starts with and then goes beyond cultivating appreciation. Studying the biosphere entails rigorous *cognitive* demands.

The purpose of this chapter is to consider the cognitive challenges involved in learning to interpret the biosphere. I organize biospheric perception into four cognitive categories, based on the juxtaposition of place and time, using the prefix (*inter-*) as a guide—interspatial, interspecies, intertemporal, and intergenerational. The prefix *inter-* has a revealing etymology. Derived from the Latin and represented in French by *entre,* it was originally used as a prefix in conjunction with the senses, denoting the ideas "between" or "in the midst." In the sixteenth century it was more widely extended to include a mutual or reciprocal quality (interdependence) as well as further elaborations of the idea "between"—including "among," "forming a link," and "belonging in common to." It's a widely used prefix because it has such conceptual power.

We need lots of ways to express the relationships between things, a means of describing how what seemingly discrete ideas, objects, and actions have in common. It is a particularly apt prefix for ecological thinking and for interpreting the biosphere. The juxtaposition of scale is based on learning how to move between and how to connect diverse conceptual worlds. I imagine *inter-* as the linguistic inspiration for a biospheric curriculum. In the next few sections, I apply the prefix to place, species, time, and generation, and consider the educational implications of doing so.

These four realms serve to synthesize and reorganize many of the curricular ideas scattered throughout the book. They are not meant as specific guidelines as much as a catalog of ideas to explore. If you're an educator, perhaps there are some activities and approaches for you to experiment with. For the casual reader, here are some ways to reflect on how you learn about the biosphere. For researchers, there are dozens of learning relationships to investigate. We know very little about the cognitive origins of ecological learning and biospheric perception. Possibly, you're reading this book with all three roles in mind—it's the combination of these approaches that yields the most insight.

Throughout the chapter, I suggest that there are forms of "pattern learning" in each of these domains, familiar processes such as structures, cycles, and relationships that serve as conceptual guides. Further, I consider whether there are appropriate developmental sequences for learning these patterns, various stages in peoples' lives when they are most conceptually ready to engage in such exploration. This is a great frontier of research for biospheric perception—merging the conceptual challenges of global change science with human cognitive development. A successful biospheric curriculum requires not only the finest global change science, but also an awareness of how people learn to perceive place and time, and what educational methods can serve both purposes. At the conclusion of the chapter, I suggest that there's an ethical foundation at the core of biospheric perception—how does a person move from appreciation to reverence?

In the last few hours, gusts of wind from the southwest have been whooshing through the trees, beckoning me to listen, drawing my attention from the writing pad. These winds are the space in between weather systems. Pulled by a high-pressure system to the north, moist Gulf air is quickly spreading into New England, carrying warblers, butterflies, seeds, and spores. The wind carries spring on its shoulders. And with spring it carries a message of hope, pronouncing its arrival, making itself available to all who will pay attention, demonstrating again that the biosphere always surrounds you, and learning its ways is a birthright and responsibility. How hard can it be to learn about something that you've always known? The sound of the wind is the music of the biosphere, and within it lies a symphony of perception.

Interspatial

Interspatial refers to the relationship between places. The ideas "local" and "global" are essentially guidelines for interpreting global environmental change, with the implication that it's helpful to sort out a range of interspatial issues. From my perspective, local is simply the bounded place from which you perceive the world at any given point in space and time. Many locals added together, compared, and then generalized suggest global. The issue isn't so much that all places are connected (one of the great clichés of modern environmental studies), as it is understanding which connections are most important. To interpret global environmental change, you have to think about the salient connections between places—the movement of people and species, weather systems, and biogeochemical cycles. What spaces between places should you observe? What is the relationship between regions? This is the "interspatial" learning challenge of biospheric perception and this section shows some of the ways that this challenge may be approached.

An interesting way to think about interspatial relationships is to conceive of your local place as a bunch of interconnected miniregions. Here are several ways to do this. Divide any small topographical region (a low hill or a gully) into two smaller regions, a north and south side. Immediately you will notice subtle changes in vegetation, determined by microclimatic nuances. Further divide those regions according to other topographic features (watercourses, rock outcrops, human settlements, soil types, etc.). On most terrain, with careful observation, you can delineate dozens of miniregions within ten to one hundred acres. What is different about these miniregions and what do they have in common?

Many interesting spatial delineations can be constructed within walking distance of one's home. In hilly or mountainous areas, it's instructive to compare elevations—what are the vegetation changes as you ascend and descend? If you live in a flat area, find the geographic boundaries that create minihabitats. On forest hammocks in Florida, habitats may change for every one foot of elevation depending on water flow, soil properties and microtopographic features. Explore a one-mile section of a watershed and observe the impact of upstream events downstream. Study the ecology of a highway system, both in terms of how the highway creates a boundary between some habitats and how it connects others. Or observe the different minienvironments that form at various spatial intervals from the road.

In chapters 4 and 5, I suggested a range of observational activities that promote interspatial thinking—tracing edges, moving from the city to the countryside and back again, or following a weather system around the world. This contributes to a biospheric perspective when you begin to notice the similarity between landscape patterns at different scales, or what might be described as "interspatial pattern learning."

For example, landscape ecologists describe watercourses and landscapes as "nested hierarchies" in which specific spatial arrangements occur at multiple scales. Think about how water moves through a landscape. Depending on topography and flow rate, water will display various patterns that transcend scale. Brooks, streams and rivers may meander, branch, or form dendritic structures. Similarly, one can observe landscape patterns at various scales—patches, corridors, and mosaics. Ecologists study these patterns to interpret how organisms live and move on a landscape. This colorful, visual, spatial conceptual language provides a way to think about the relationship between places,

regardless of their size.³ *Similar perceptual tools can be used in different eco-*
logical regions and landscapes. This pattern language, grounded in natural
history and biospheric phenomena, is experiential and applied—you
can see these principles at work anywhere.

Wilson and MacArthur's theory of island biogeography uses interspatial pattern learning to interpret biodiversity. The "distance effect," describes the tendency of remote islands to contain fewer species than
islands that are closer to other islands or to the mainland. The "area effect" reflects the tendency of smaller islands to have fewer species than
larger islands.⁴ By studying the relationship between area and distance,
ecologists learn how geographic features (spatial patterns) promote or
restrict biodiversity.

The study of island biogeography is currently receiving increased research attention because ecologists and conservation biologists recognize that urban development creates island effects, or *patches,* in
land-based regions. Buildings and highways separate habitats. This creates landscape patches that resemble islands. Hence the species-area relationships observed by Wilson and MacArthur may be relevant to a
variety of ecological and geographic settings.

Landscape ecologists consider the ecology and morphology of
patches—their size, shape, and number; their boundaries and edges. For
example, *corridors* are areas that link patches. Conservation biologists
study corridors to determine whether species can move unimpeded between patches. Different types of corridors have different attributes,
such as topographic and ecological characteristics. Roads, power lines,
windbreaks, woodlands, streams, and rivers all may serve as corridors.
The size and composition of a corridor and its relationship to a patch are
of great interest for biodiversity management because they help assess
the viability of species survival when a habitat is threatened. Patches and
corridors form land mosaics at different scales. The interpretation of
land mosaics is an important tool for biogeographers. A landscape ecology textbook reveals a detailed lexicon of patterns, a pattern language
that provides the user with the flexibility to shift through variable scalar
hierarchies. You can use these spatial methods at the intimate scale of a
proximate community, or you can look at an entire biome this way.

Philip Ball's book, *The Self-Made Tapestry: Pattern Formation in Nature,*
explores universal patterns, found at different scales in nature. A chapter on "Branches" shows how branching patterns emerge in snowflakes,
flowers, trees, and rivers. Ball shows that "we can recognize qualita-
tively similar branching patterns in a huge variety of different physical

and organic systems . . . this leads us to the compelling conclusion that there is indeed something generic—something universal—about these forms, and by extension about the rules for their formation."[5] Another fine example of interspatial pattern thinking is Philip and Phylis Morison's inspirational *Powers of Ten,* a mind-blowing journey through exponential scale. Starting with a man sitting on a blanket, having a picnic in a park near Lake Michigan, the reader is treated to a series of photographs and drawings corresponding to what you would see at different magnifications, both micro- and macroscopically—from the infinitely large to the infinitely small and back again—all starting from your patch of picnic ground. There are few better illustrations of size, scale, and distance.[6] Finally, Alexander, Ishikawa, and Silverstein's classic planning text, *A Pattern Language,* develops a methodology of interspatial thinking as a basis for ecologically sound town and regional planning.[7]

In browsing these texts, you are deeply moved both by the ecology of pattern and the sheer joy of aesthetic expression. Interspatial pattern learning is both conceptual and experiential. Moreover, it is rewarding to engage in. Yet I'm struck by the relative absence of this kind of pattern learning in most school settings. One can imagine a provocative and engaging curriculum built around place-based learning that develops the cognitive skills of interspatial thinking. What a fine means to link ecology, literature, and art. Observing and experimenting with different views and scales of a landscape allows a student to viscerally experience the relationships between places. Such movement between spatial worlds spurs storytelling (exploring scale is a journey) and artistic expression (you may be moved to draw, photograph, or compose music about these patterns).

Most of these examples of "interspatial pattern learning" are ways of modeling how exemplary biospheric naturalists and scientists observe the world, and then emulating their methods. What is most relevant to biospheric perception is noting the specific cognitive tools that these naturalists use and then applying them in diverse educational settings. What educational preparation and special talent or insight allows a person to develop such facility at pattern learning? What are the specific qualities that allow a person to explore the "inter" learning zone?

Consider an interspatial concept like the hydrological cycle. It takes a good deal of scientific training to understand the movement of water—evaporation, precipitation, condensation, fluvial geomorphology (drainage, stream flow, floods). If the hydrological cycle were taught as an exercise in shifting scales, what experiments, activities, artwork, and stories would serve as a sound learning sequence? What tangible experiences might be organized? How would you handle, explore, and play with water? Perhaps different aspects of the water cycle could be explained at different cognitive stages. This could be reinforced through the continuity of teaching the same material (but at different conceptual levels) at each stage of schooling. The same educational strategy might be developed for all the biogeochemical cycles—carbon, sulfur, nitrogen, phosphorus, and so on. Understanding them in depth, both viscerally and intellectually, would forever transform one's interspatial perceptual faculties.

Thinking about interspatial pattern learning opens up a whole new realm for curriculum design. In planning a biospheric curriculum, inevitably you think about the specific content to be covered—what mix of natural history, ecology, earth systems science, and conservation biology should be offered? Equally important is the consideration of the developmental sequence corresponding to various stages of pattern learning. Developmental theorists have researched all kinds of sequences for math, reading, writing, and traditional academic skills. But might there also be a developmentally based sequence for learning about ecological relationships, global environmental change, and the biosphere?

I am suggesting that the process of learning about global environmental change—the ways a person learns to experiment with scale, to broaden conceptions of space and time, to identify with other species, to consider posterity—are as important as the subject matter. With a few exceptions (Paul Shepard, Joseph Chilton Pearce, Edith Cobb, David Sobel, Louise Chawla, and Peter Kahn) educators have paid very little attention to the developmental origins of ecological learning. What are the special cognitive qualities that allow people to think in (ecologically) interspatial and intertemporal terms and how might they be taught? Are there specific developmental stages in which these qualities are more accessible? At what ages are children and adults receptive to learning various biospheric concepts? What are the field-based observational activities that enhance such perceptual abilities?

We need "ecological Piagets," researchers and theorists who combine ecological training and biospheric thinking with an understanding of developmental psychology. Such research might involve (1) specific, field-based experiments that attach cognitive thinking to ecological knowledge, (2) probing qualitative interviews with ecologists and earth systems scientists to ascertain how they learn to think in biospheric terms, and (3) collaborations with indigenous cultures for whom such thinking is intrinsic to survival.

During middle childhood, I took great pleasure in exploring various ways to divert water. At the beach, I would enjoy finding inlets and sandbars. What routes and courses might emerge from building little channels and runnels? Or if someone in the neighborhood were washing the car, the local kids would find ways of diverting the runoff as it careened down driveways, sidewalks, and curbs. I rushed to join them. These were great lessons in interspatial pattern learning and with just a little bit of prompting, I could have learned a great deal about geomorphology, the water cycle, and branching patterns. Good elementary science teachers know how to take such experiences and turn them into ecology lessons. To organize a biospheric curriculum we need dozens of such approaches, geared to the appropriate developmental sequence, linked to interpreting global environmental change. And we need good research programs to investigate the relationships between pattern learning and ecological knowledge.

Interspecies

After several days of mild, southwesterly wind, on May 5, spring is gushing forth like a flood. This morning, the forest is filled with noise. There are dozens of bird calls. The year-round residents are busy and active—chickadees, nuthatches, woodpeckers, and titmice are scurrying around, frenziedly feeding, calling from what seems like every perch. There are lots of migrants moving through. A white-throated sparrow is singing, on its way to a boreal forest somewhere north of here. I can hear, but not quite see, the common, early migratory warblers—black-throated green, black and white, and ovenbirds. Lots of wildlife is on the move. There have been several reports of moose-caused motorcycle crashes. The beavers are very active, the black flies are swarming and biting, there are still plenty of spring peepers, and a few weeks from now turtles will begin to lay their eggs.

Between all of these tracks, signs, songs, scats, and behaviors, a great deal of information about the ecosystem is revealed, much more than I can possibly observe. Whenever I sit silently and listen to the depth and richness of animal sounds, it occurs to me that most of these creatures are far more sensitive to my activities than I am to theirs. By studying the life history and listening to the stories of any of these species in detail, I could learn a great deal about not only this forest habitat but the state of the world as well. Each of these critters broadcasts the news of the ecosystem.

Richard Nelson, in his classic study and testimony to the Koyukon people of northwest Alaska, *Make Prayers to the Raven,* shows how for the Koyukon, the origins of ecological learning flow from species stories. "These stories constitute an oral history of the Koyukon people and their environment, beginning in an age before the present order of existence was established. During this age 'the animals were human'—that is, they had human form, they lived in a human society, and they spoke human (Koyukon) language. At some point in the Distant Time certain humans died and were transformed into animal or plant beings, the species that inhabit Koyukon country today. These dreamlike metamorphoses left a residue of human qualities and personality traits in the north-woods creatures."[8]

Both through the perennial retelling of the stories of the Distant Time, and the full immersion in subsistence living (and the natural history observation that goes with it), "there are hundreds of stories explaining the behavior and appearance of living things."[9] Much of Nelson's book explains the Koyukon relationship with plants and animals, their extraordinary knowledge of their natural history and ecology, and the lively and revealing narratives they inspire. Animal and plant names are a blend of onomatopoeic sounds and salient life history characteristics. Each species carries both a wealth of ecological learning and an encyclopedia of earth wisdom lessons.

Nelson describes a form of traditional "interspecies pattern learning"—a combination of story, subsistence, and natural history that relies on the interpenetration of humans, animals, and plants, stemming from a time, when they all spoke a similar language. Throughout most of human history, this is the means through which environmental education has occured. Sometimes shamanically, but also through the day-to-day encounters of different species, one learns how to travel through "species spaces" as a means of learning about the ecosystem.

Paul Shepard suggests that the interpenetration of human and plant and animal lives is crucial to personal and cultural development. "The discovery of the external existence of animals who correspond to our inner reality began not as an invention but in the natural history and evolution of mind."[10] Shepard emphasizes that human contact with the "wild other" is crucial for ecological perception. The stories, metaphors, and rituals of interspecies learning are the developmental foundation for ecological identity. Shepard fears that "the loss of the wild Others leaves nothing but our own image to explain ourselves—hence empty psychic space."[11] His brilliant commentary on interspecies learning is aptly summarized here:

Cognitive taxonomy and artifacts are indeed the tools in the perceptual work by which the whole person is achieved. But the effect of a healthy identity and maturity is realized in attitudes towards the environment, a sense of gratitude more than mastery, participation in a rich community of organisms, a true biophilia or polytheism. The images—animal guides and mediators—are the representations of an outer world that made our own being possible and toward which our maturity has its end: the preservation of the world. The obligations of having evolved in natural communities constitute a kind of phylogenetic felicity in which we acknowledge that the fish, amphibian, mammal, and primate are still alive within us and therefore have a double existence. They are present as bits of DNA, affirming kinship, and also in the world around us as independent Others. The concept of biodiversity as a social value grows from an inner world and creates respect for a mature ecology, that is, 'climax' ecosystems with their diverse inhabitants.[12]

Why stop there? Lynn Margulis and Dorion Sagan expand Shepard's vision to include microbial life as well. In *What is Life?* (see chapter 5), they lay out the full magnificence of all five taxonomic kingdoms—bacterial, animal, plant, protists, and fungi—demonstrating the extraordinary intelligence that pervades each of these realms. Margulis and Sagan have a video series in which they portray the growth forms, movements, and reproductive behaviors of diverse microorganisms.[13] Whenever I watch these videos, I feel that I am (at least virtually) entering the wild Other of microbial species space. Knowing that these microorganisms pervade my body, the soil, and all dimensions of the biosphere evokes a kinship for sure, but it is also profoundly disconcerting as you feel the relative insignificance of humanity. You are ruthlessly put in your place. Margulis and Sagan describe such feelings of discomfort as "healthy" in that your biospheric perception is broadened through interspecies awareness, and your humanity is viscerally placed in its limited context and scale.

Recall Jacob von Uexkull and the concept of umwelt described in chapter 4. Uexkull is interested in ridding education of the misconception that all species share a similar conceptual space. We look at the world as if all creatures perceive as we do. In the introduction to "A Stroll through the World of Animals and Men: A Picture Book of Invisible Worlds," Uexkull maps out the phenomenology of umwelt, or how to interpret sensory perception in different species. Uexkull provides some remarkable perceptual guidelines (written in 1934) which are of great use for interspecies pattern learning.

He proposes a "stroll into unfamiliar worlds," a journey into "worlds strange to us but known to other creatures, manifold and varied as the animals themselves." Uexkull sets the scene—a rolling meadow, adorned with flowers, strewn with insects and butterflies. He suggests that the observer imagine a soap bubble around each creature, to indicate distinct fields of perception. "When we ourselves then step into one of these bubbles, the familiar meadow is transformed. . . . Through the bubble we see the world of the burrowing worm, of the butterfly, or of the field mouse: the world as it appears to the animals themselves, not as it appears to us."[14]

Contemporary nature literature abounds with examples of writers who aspire to cultivate interspecies learning. Freeman House follows salmon to the extent that his entire identity, his understanding of the Mattole River watershed (northern California), and all of his community organizing is wrapped in his profound love and respect for the life story of salmon. It is through salmon that House reveals the continuity of the landscape, kneading folk tradition and local natural history, developing a resilient cultural ecology that spans diverse times and places.[15] In *Heart and Blood*, Richard Nelson reports on his years following deer around North America. In his lifelong quest to study and emulate deer, he finds that his entire way of life is transformed. As much as is humanly possible, he learns to move, think, and live as deer do.[16] Scientists who research animal perception reveal entirely new dimensions of sensory experience, enhancing the senses more profoundly than any technology, disclosing realms of perceiving way beyond what humans can imagine.

How might interspecies pattern learning be addressed developmentally? The first challenge is to make sure that proximity to critters is prevalent in all educational settings. Either you bring "live visitors" into classrooms for short periods of time, or much better still, students and teachers spend their time outside, in the presence of living things. Second, it's crucial to incorporate Paul Shepard's observations regarding the symbolic significance of animals and plants for personal identity. Third, consider at what age students are most interested in exploring the life worlds of various creatures.

Young children tend to be interested in small mammals. In *Beyond Ecophobia*, David Sobel suggests that identification with animals is the most appropriate environmental education activity for early childhood (ages 4 to 7). He describes how annoyed he gets when fervent educators lay birdwatching trips on young kids, or even adolescents who may have no interest in learning the names of dozens of birds, and lack the patience to sit for hours and observe them. Sobel suggests that young children are interested in how birds make nests and create homes for themselves.[17] Here is a developmentally appropriate integration of natural history, ecology, and life story that allows kids to identify with birds, and learn from them. Similarly, during adolescence, more ritualistic relationships with animals and plants that blend adventure and ecology better speak to the needs of nature-based development. An entire developmental "interspecies pattern learning" scheme can be developed.

One might imagine a "five kingdoms" curriculum through the entire human life cycle (merging science, storytelling, and art). Studies in ethnobotany, animal behavior, microbial ecology, and fungi could be staged accordingly. This is not so far-fetched as it might sound. Paul Shepard's work is rich in such theoretical suggestions. Many traditional cultures that retain their shamanic ways have thought a great deal about these issues. It would be inspiring to gather ecologically trained developmental theorists, curriculum specialists, and indigenous elders to work with biospheric naturalists to set up a five kingdoms curriculum that ranges from ethnopoetics to microbial ecology. What guidelines and curriculum might they develop for interspecies pattern learning?

Intertemporal

Anytime you observe nature, you mark the passage of time. Whereas ten days ago it was snowing in the Monadnock forest, today (May 8) it is hot and humid. It feels like midsummer. The Canada mayflower leaves are fully unfurled and their flowers are blooming. The wild cherry and birch blossoms are leafed out. Wasps, ants, and mosquitoes join the black flies. The peas and greens are shooting out of the earth. Wild oats and violets are in bloom. I hear dozens of bird songs echoing throughout the forest.

J. T. Fraser, a philosopher of time, reminds us that "processes suitable for timekeeping are everywhere: pine cones in my study which open when ripe, geese migrating every spring and fall in response to some ancient call, the sun rising with great probability every morning, and the hum of my Bulova Accutron doing its regular $1.136,003,398,424 \times 10$

(10) cycles per sidereal year."[18] Yet with the exception of the most obvious seasonal observations, most people mark time with personal accomplishments and milestones, deaths and births, and the body's physiology—time's most tangible marker is your gray hair. We pay more attention to the passing of digits on the hands of a watch than we do to subtle changes in the landscape, migration schedules, or the sprouting of seeds. Nevertheless, our physiology demands that we pay attention to the biology of time, and there is no more tangible way to do so than watching yourself age in comparison to what you observe in nature.

For biospheric perception, the great challenge is how to teach ideas of time using natural history and global environmental change as a guide. How do people learn to travel through time, to view biospheric processes with either a single day or a hundred million years as a context? The significance of issues like species extinction and global warming become most apparent when you use biospheric time as a criterion for change. Are there approaches to "intertemporal pattern learning" that enhance biospheric perception?

Many of the world's great meditation traditions emphasize the importance of following the breath as a means to cultivate awareness of the present moment. This is an appropriate guideline for a biospheric curriculum of intertemporal learning. In chapters 4 and 6, I explained how the deliberate gaze of the naturalist leads to place-based paces, building one's perceptual alacrity, providing a focus and foundation for more far-ranging explorations. You find your center when you slow down. One of the extraordinary qualities of time is that when you fully arrive in the present, the full spectrum of past, present, and future is more easily perceived. This may be a useful approach for intertemporal pattern learning—the ability to ascertain patterns as you compare various time frames with what you observe in the present.

Previously I suggested that a good way to think about spatial relationships is to divide a place you know well into bunches of miniregions and then see how they connect to each other. The same process can be used for intertemporal learning. Over the course of a week during autumn, notice how the leaves change color on the trees. In springtime, observe how much time it takes for a flower to bud and bloom. During the course of twenty-four hours, there are innumerable cycles based on the alignment of the earth, sun, and moon, the movement of weather systems, or the arrival and departures of people and species.

Another concrete approach is to trace the origins, lineage and ancestry of people, species, and landscapes. You start by considering how old things are. How old are turtles, trees, boulders, lakes? How did they come to be in a given place? The most tangible way to understand the age of things is to learn how to track the evidence of time. Although some instrumentation is required, there are some hands-on, field-based paleoecological techniques that allow you to reconstruct Pleistocene environments. In *The Holocene: An Environmental History,* Neil Roberts explains many of these techniques. Dendrochronology (counting tree rings), lichenometry (measuring the size of lichens), and sedimentology (measuring the size and grain of sand and pebble deposits) are excellent place-based techniques for observing recent geological time.[19] Martin Lockley, in *The Eternal Trail: A Tracker Looks at Evolution,* uses fossil footprints both as a means to understand geological time, and also as a way to study animal behavior and ecology. Through hands-on, participatory paleontology, the expanses of time are reduced to tracks you can see— visceral links between eras.[20]

Tom Wessels (see chapter 4) shows readers how to observe the history of environmental change in the forest. By looking carefully at signs such as stone walls, tree stumps, boulders, streambeds, and other topographic features, Wessels reconstructs recent human and ecological history (three hundred years' worth). With some additional training in geomorphology, you can go back twenty thousand years or so, reading the lakebeds, hillsides, and waterflows for signs of glacial deposition and scouring. Travel further still with a good geologist who can show you how to interpret rock formations and help you understand what a million years means. A paleontologist will use fossils to accomplish this. And an evolutionary microbial ecologist will sample some water from the neighboring beaver pond, stick it under a microscope and take you to the earliest days of life on earth. It takes the actual experience of doing these things to begin to get a grip on evolutionary time. All of these scientists practice "intertemporal pattern learning" by virtue of such investigation.

Barbara Adam, in *Timescapes of Modernity,* shows how global environmental change is difficult to perceive because many of its processes are linked to the invisibility of time. Landscapes as chronicles of "life and dwelling" are more readily perceived than "timescapes" which involve a "shift in emphasis not just from space to time but, more importantly to that which is invisible and outside the capacity of our senses."[21] Adam suggests that we cultivate a timescape perspective. "Where other scapes such as landscapes, cityscapes, and seascapes mark the spatial features of past and present activities and interactions of organisms and matter, timescapes emphasize their rhythmicities, their timings and tempos, their changes and contingencies."[22]

Very little work has been done among environmental educators to explore the developmental sequence of learning how to perceive time. Given the importance of time perception for interpreting global environmental change, such research is crucial. Is the ability to think in specific temporal intervals linked to cognitive development? What are the implications for teaching evolution, ecology, and biogeochemistry? At what age is a student ready to consider the geological time scale? And what patterns must he or she explore to make such learning interesting and relevant?

It wasn't until midlife that I began to take an interest in the history of life on earth. This interest was prompted by a spiritual search to understand the origins of humanity. And I still find it extremely difficult to grasp the geological time scale. In all of my formal schooling, I received virtually no instruction in how to think about time. Usually, it is only scientific specialists who learn how to do this. If this is true for contemporary education, how can we expect people to understand global environmental change? A biospheric curriculum must consider time cognition and its relationship to environmental learning. How ironic that there are shelves of popular management books that instruct people and organizations how to "manage" their time, yet most people pay little attention to the concept of "deep time," which is the ultimate context for life on earth. A prevailing challenge for biospheric learning is how to bring awareness of deep time into educational settings.

There are some interesting efforts to do so. Stewart Brand's book, *The Clock of the Long Now,* aspires to create an institutional and cultural context for an expanded notion of time by calling public attention to an actual clock that measures for the present and posterity simultaneously. Thomas Berry, Brian Swimme, and other theologians of various persuasions are reconstructing creation myths that incorporate ecology, evolution, geology, and cosmology, thereby calling attention to the meaning of time. Scientists such as Lynn Margulis are making concerted efforts to develop curricula that probe the history of life on earth.

Every educational hunch and experience I have leads me to assert that time, species, and places are best studied together, through a combination of detailed observations of natural history and the reassurance of cultural and historical continuity. The ability to perceive time in nature is effectively grounded by the legacy of human kinship and cultural tradition. Time cognition has an important analytical component, but it's also deeply intuitive—time is measured by both the head and the heart. Perhaps "intertemporal pattern learning" is also taught through the wisdom stories that multiple generations collectively weave.

Intergenerational

It is rare in contemporary educational settings for elders and teenagers to exchange views about the issues that are most important to them. People in the same age group tend to stick together. Often, authority issues, family stereotypes, and cultural differences encumber intergenerational relationships. Elders wonder whether anyone is listening to them. And young people desperately need to step out of their youth culture to speak with folks who have been around for awhile. This overriding sociological issue is directly relevant to environmental learning as well, where younger people are overwhelmed by the magnitude of their concerns, thinking that they are unique to their generation.

In October 1999, I became involved with a group of students, teachers, professors, and writers who organized an intergenerational forum on environmental activism. A circle of elders, drawn from diverse ethnic and conceptual backgrounds, gathered with aspiring environmental professionals and service-oriented high-school students to exchange ideas about legacy and posterity. Through a series of panel discussions, small-group workshops, poetry, storytelling, artwork, and ceremony, the elders conveyed stories of hope and environmental activism. The younger students had a venue to express their concerns and motivations regarding the future of environmental issues. The forum was designed to provide a younger generation of activists with a sense of legacy for their work. Also, in the spirit of educational experimentation, we wanted to see what would happen when several generations would have an opportunity to openly share their concerns and aspirations about pressing environmental issues.

As preparation for the event, each of the high-school students corresponded with one of the elders, based on common interests. Each student chose a mentor accordingly and wrote a note introducing herself and explaining what she was interested in learning. These students then joined with an elder in an intimate workshop setting. A naturalist led a drawing workshop. A rabbi led a session on prayer and ritual. A bioregionalist designed a mapping activity. A civil rights activist ran a small workshop on political action. The forum worked out extremely well, including discussions of global environmental change, issues of hope and legacy, and dealing with various moral and ethical dilemmas.

As the day-long intergenerational forum drew to a close, the participants were standing in a circle on a grassy knoll, surrounding a tree planted earlier in the day by Jake Swamp (a Mohawk environmental activist) and a half-dozen students from a local high school. It was a mild October evening. High pink cirrus clouds signaled the first phase of dusk. The leaves were golden brown and orange, the last vestige of autumn brilliance. The subtle hues of the sky and trees produced a calming effect, a nice touch to a day filled with so much authentic discussion and opened hearts.

Jake Swamp sang a Mohawk prayer, giving thanks to the day by virtue of the spirit of the land and its ancestors. As he finished there was a prolonged silence. Feeling somewhat responsible for getting everyone home on time, I thanked people for their fine participation and called an end to the day. But this pronouncement was unwarranted and unnecessary. The circle wouldn't dissipate. It became clear that no one wanted to leave! The power of the day was so resounding that people had no idea what to do next. The forum had unleashed so many ideas and so much good feeling that closure seemed impossible. Rather, people felt that their discussions kindled hope and they were enormously grateful for the authentic talking and listening. The participants were bearing witness to this process and they were unwilling to let it go. People spontaneously began to thank each other for the experience. When sufficient thanks were bequeathed, the group's momentum gently and gradually shifted pace, and the forum faded into the next phase of dusk.

Later that evening, the elders and staff had dinner together, still basking in the glow of good feeling, absolutely enamoured with the moving and penetrating comments of the students. What happened that made this day so powerful? What were the qualities of the experience that made it so genuine? Why did all of the participants feel that they were engaged in such an inspirational educational experience? Two weeks later, I facilitated a short seminar at the local high school so the participating students and teachers could talk further about the meaning of the day. Here are the key themes that emerged.

First, high-school students crave discussions of moral and ethical issues. The students all expressed how much they enjoyed talking about their hopes and concerns regarding environmental questions, whether it involved difficult, existential, and spiritual issues such as the meaning and purpose of human life, or more tangible feelings about their relationship to the land or their sense of place. They mentioned that it is almost impossible to have such discussions in the school atmosphere, both because it doesn't appear to be relevant to the curriculum, and because they feared how their peers would perceive them if they ventured into such territory. The intergenerational forum gave them a "safe place" to discuss such issues, in an out-of-school setting, where there were structures in place to support such discussion. Students who otherwise would never even speak to each other (by virtue of the tight and selective social pressures in a high-school setting) found that they had much in common. They all agreed that schooling would be much more relevant, interesting, and profound if there were more venues for discussing their deepest moral and ethical concerns, in this case, regarding their relationship to nature.

Second, high-school students and elders are often marginalized in public discussions of moral and ethical issues. Many of the students expected that when they attended the forum, they would just be "talked at" by the conference organizers and the elders. They are used to situations in which their opinions are ostensibly solicited, but in which they wind up being passive observers. They were genuinely delighted and surprised when they realized that there was great interest in what they had to say. One student poignantly suggested that it was middle-aged people who think they know everything. It's middle-aged people who make all of the decisions, and who ignore advice from young and old alike. The elders, he suggested, were secure enough in their knowledge and experience that they didn't have to prove anything, hence they were more open and engaging to speak with. Some of the students also expressed their disappointment at having few "elder role models" in their lives. All of the students expressed their keen desire to work with elders more frequently, to find ways to do this in both school and community settings.

Third, high-school students have a tremendous capacity and willingness to engage in civic and environmental service, but they need a context for their work. Few of the participating students were aware of the legacy of activism that they inherit. It is one thing to read about civic heroes in a textbook. It is another thing altogether to have direct encounters with people who have dedicated their lives to service and activism. The strongest context for such work is the transmission of stories. Elders require milieus in which they can tell such stories and high-school students need to hear them. Yet it was also evident that the high-school students have their own stories to tell. Having an elder with a distinguished career of service and activism listen to your stories is an extraordinary honor. In such situations, you find common aspirations that transcend your era, but take on a century of historical experience and a range of geographic and ethnic backgrounds.

Fourth, elders are inspired and invigorated when working together with youth. The authenticity and openness of the high-school students moved the participating elders. The conceptual insights and perspectives that the students brought to the discussions delighted them. Just as the students require a context for their interests, so do the elders need the perspectives of youth to lend a sense of posterity and legacy to their work. The two-way transmission of stories allows for the passing of wisdom. "Working together" implies sharing skills and rituals as well as life experiences.

"Intergenerational," or moving between generations, provides a human context for teaching about global environmental change. The most evocative lessons emerge in the course of daily contact with knowledgeable mentors—people who can show how you how to do something, tell you about what happened to them, take you under their wing, and show you that they care about what you do. Likewise, mentors need inspiration from their students, the sense that their teachings have continuity, and will be reinterpreted by another generation.

It's revealing for young and old alike to witness how the world looks to each other, or to imagine how the world changes through time. My grandparents' childhood in the Russian pale is a remote dream, as far removed from me as my own children's old age. Yet there is continuity between these three generations of history and tradition, lodged in the landscapes of their lives through their common hopes and aspirations. In a world of perennial change, that legacy endures.

Interpreting the biosphere is a community effort, requiring full engagement by diverse cultures and multiple generations. What better record of the recent past can there be than the naturalist musings of the farmer or woodsman who lived in your place before you were born? Discussions of biodiversity are enhanced when young and old compare what they see in the landscape, so "generational amnesia" (see chapter 4) can be avoided. I imagine grassroots environmental change workshops that gather young and old, members of diverse cultures, experts and amateurs, ecologists and artists, all sharing their impressions of the living landscape through field natural history, artwork, memoirs, and spiritual concerns.

This intergenerational learning doesn't require any special developmental curriculum—all of the age groups are there working together. They figure out through common sense and plain talk just what can and can't be understood. The human life cycle meets the biosphere through direct contact. The good teacher brings these groups together through the crucible of community concern. A biospheric curriculum breaks down the barriers between school and community, and turns the school into a living laboratory of mutligenerational vitality.

As a fifty-year-old basketball player, I derive incredible pleasure from playing hoops with high-school players. Sometimes the older guys take on the young players and we are victorious by virtue of our experience and knowledge. And sometimes the younger players figure it out and through their energy and imagination they teach us a thing or two about the game. On other occasions we all mix together, building community through the sheer joy of playing something that we all love. I've made connections through the authenticity of basketball that would have taken me years to make in other contexts.

The same type of intergenerational play should be a foundation for schooling—young and old taking delight in the world together. Interpreting global environmental change, as serious as it is, shouldn't only be heavy and deep. There is a lightness and improvisational quality to loving the world, and young and old need each other to make this happen.

There is much talk about family values and spiritual tradition in American life. I'd like to see biospheric learning serve to reinvigorate family life—multiple generations exploring the world together as a means of bonding family and community. Why not expand the boundaries of family so as to include the "more than human" community and the diverse cultures of people and species living together? Inevitably there will be conflicts and differences, but no more so than you typically see in any homogeneous family. Let the natural world be the place where such differences are encountered and explored.

On the ground level, when people and species share what it means to live in common on a landscape, they develop affiliations with habitat and community. These affiliations are delivered through art and story, based on the intimacy of shared knowledge. When you experience such affiliation, and do so with your friends and neighbors, wherever they may reside, whether in close proximity or just passing through, you cultivate empathy and compassion. You begin to bring the biosphere home, not only on the wings of perception but by opening your heart.

Curriculum as Mitzvah

A place-based perceptual ecology has a twofold purpose. Through familiarity and intimacy, you learn how to pay closer attention to the full splendor of the biosphere as it is revealed to you in the local ecosystem. In those moments when you can wade through the distractions of business and task, when you catch a glimpse of the unfathomable world at your doorstep, you open yourself to biospheric perception. Through a deliberate place-based gaze, by learning how to move between worlds, you allow those glimpses to last a little bit longer each time. By developing appreciation for the biosphere, in liberating your sense of wonder, in summoning praise and reverence, in contemplating the mystery and circumstances of processes that you can never fully understand, you feel a sense of gratitude and appreciation. You learn to honor biogeochemical cycles as intrinsic to your breath and thirst. You find your origins in the history of life on earth. You forge alliances and affiliations with people and species from all corners of the globe as you watch them pass through your neighborhood. You summon praise for whatever lies behind this outstanding journey—Gaia, God, evolution? With the passing of praise comes cause for celebration.

There is also great risk in studying the biosphere, synonymous with the very peril of living—the danger of swimming perennially in a sea of insignificance, wallowing uselessly in a world seemingly out of control, beyond your understanding, without continuity or purpose. "What is the truth of being human?" asks Heschel (see chapter 3). "The lack of pretension, the acknowledgment of opaqueness, shortsightedness, inadequacy. But truth also demands rising, striving, for the goal is both within and beyond us. The truth of being human is gratitude; its secret is appreciation."[23]

At the core of a biospheric curriculum is a set of qualities—wonder, indebtedness, appreciation, gratitude, praise, and reverence. These qualities aren't taught as much as practiced. To a certain extent every person has to learn for himself how to exemplify these qualities. You cannot structure intimacy with the biosphere. There are occasions, moments of great awareness and serendipity, when you feel that you are deeply touched by something unfathomable, far greater than you, beyond contemplation, as you view a sunset, watch a thunderstorm, listen to a bird call, or dig in your garden dirt. What a biospheric curriculum can provide is a forum for discussing these moments, a way to make them meaningful in the context of the human community, a means for supporting and challenging one's impressions and considering their moral implications. What is most crucial is providing a way for a person to act on her impressions, to allow for the expression of gratitude and celebration in the form of tangible deeds.

Here lies the ultimate purpose of schooling, whether in the classroom, the field, the family, or the spiritual center. Again, consider Heschel: "It is a most significant fact that man is not sufficient to himself, that life is not meaningful to him unless it is serving an end beyond itself, unless it is of value to someone else . . . Sophisticated thinking may enable man to feign his being sufficient to himself. Yet the way to insanity is paved with such illusions. The feeling of futility that comes with the sense of being useless, of not being needed in the world, is the most common cause of psychoneurosis. The only way to avoid despair is *to be a need* rather than an end. *Happiness,* in fact, may be defined as the *certainty of being needed.* But *who* is in need of man?"[24]

Heschel's entire theology is built around this profound question. Who needs you? Your mother, father, brother, dog, and cat? Your spouse and children? Your friends and associates? Your clubs and associations? Your town? Your country? The local ecosystem? The biosphere? Gaia? God? Perhaps they all do. You're needed everywhere. But what is that you're being asked to do?

At the heart of Judaism is the concept of the mitzvah. In Buddhism there are the Five Wonderful Precepts. Every wisdom tradition has an equivalent concept as rooted in its specific historical and cultural traditions. Here, I will briefly elaborate on the meaning of mitzvah and its relevance to a biospheric curriculum. In chapter 3, I cited Heschel's comment that "daily wonder requires daily ritual." He adds that "we must keep alive the sense of wonder through deeds of wonder . . . A sacred deed is where earth and heaven meet."[25]

A mitzvah is a "well of emergent meaning" in which a person expresses gratitude and praise through a concrete action in community. A mitzvah is "a deed in the form of a prayer." It's a means of serving a purpose beyond yourself, in pursuit of a greater good, a contribution to the well-being of the universe. Yet the act itself needn't be grandiose. A

profound deed takes place anywhere, wherever compassion, selfless-
ness, and reverence might emerge.[26]

For the purposes of this book, and its environmental imperative, a
mitzvah takes the form of ecological restoration. "Our task is to cleanse,
illumine, repair."[27] In Judaism, there are specific moral guidelines for
such actions, to provide a tangible context for their setting. These guidelines are always subjected to reinterpretation in the light of historical circumstances. Arthur Green, in his inspiring book, *Seek My Face, Speak My
Name*, develops a list of mitzvahs that have direct ecological application.
His fourth mitzvah, "acting with concern for the healthy survival of creation itself," suggests that "a commitment to preserving the earth also
means a commitment to preserving the great and wondrous variety of
life species in which the One is manifest."[28] In *For a Future to be Possible*,
Thich Nhat Hanh and an assortment of interdenominational commenta-
tors, consider both the ecological and spiritual application of the Five
Wonderful Precepts. The first precept is reverence for life. "Aware of the
suffering caused by the destruction of life, I vow to cultivate compassion and learn ways to protect the lives of people, animals, plants, and
minerals."[29]

Mitzvahs, precepts, commandments, "distant time" stories—these
are the converging wisdom traditions that form the basis of human action. Within all of these traditions, as applied to local ecological circumstances, there's a global encyclopedia of deeds. These represent an
ecosystem of ethical tradition, linked together through a common
awareness, that human indebtedness to the biosphere is most tangibly
expressed through the inspired actions of individuals and communities.
To my mind, a biospheric curriculum must have this ethical substrate.
Through inspired deeds you gain membership in the convergence of
these ethical traditions. Call it what you will—service learning, ecological restoration, spiritual practice—this is the glue that binds a biospheric
curriculum to the earth's wisdom traditions.

Song of Songs

Our spring heat wave finally passed. In only four hours the temperature
dropped from ninety-two to fifty degrees. And it never even rained! A
"backdoor" cold front moved from the northeast off the ocean, temporarily pushing the southern heat back where it belongs. The next day
was filled with cold rain, punctuated by spring lightning and thunder.
Today, May 11, the air is pungent and fresh. A snappy northwest breeze

jiggles the leaves. Leftover moisture drips through the canopy, adding a light percussive tapping to the sounds of the morning. The wet forest floor oozes with fragrance—a soothing mixture of pine, moss, and mold—mixing with the ambient flowery scent of tree blossoms circulating throughout the woods. The leaves are still on the yellowy side of green, but they are almost full size and you can barely see through the canopy to the sky. It's impossible to find the ovenbird chattering away in close proximity. The spring greens in the garden have grown enough in the last few days to warrant thinning. These are just a few of the more obvious changes. Much more is happening still, just beyond my perceptual reach.

Despite all of my aspirations and wishes, I'm aware that biospheric perception is as much an ideal as a practice—making sensory demands that I can't ever fulfill. It's challenge enough to keep aware in this place and time, to pay attention to the most common patterns of this fine spring day. I must admit to feeling overwhelmed, daunted by the "pattern learning" of a biospheric curriculum—traveling through multiple worlds is both thrilling and exhausting.

This past weekend I flew overnight to Denver (to talk about this material) and returned the next evening. From the plane I viewed phenomenal cloudscapes, observing the weather from above rather than below. After about ten minutes I had to look away. There was just too much to take in. While driving to my hotel at dusk, I was besieged by patterns—the glimpses of the snow-covered Rockies, the dual sprouting of spring and development (buildings are shooting up like weeds all over Denver), the beautiful ethnic mix of Hispanic, Asian, White, and Native American cultures that I never see in New England.

The next morning I went for a walk through a modest residential area, looking for signs of the landscape. I stared at the Front Range and then turned around looking east to the flat expanse of suburb encroaching on the prairie. I pondered all the field guides to "x East of the Rockies" and realized that I was standing on that boundary. I searched for difference—and even in the homogeneity of American popular culture, here was a place dramatically different from New England. The color and angle of the light, the fragrances, the expansive sky, the red pebbles, the magpies, the brick houses, the culverts, the vegetation. I was tempted to wander both west and east, beckoned by mountains and prairies. But responsibility prevailed and I returned to the hotel so I could meet my new colleagues and proceed to the event. I flew back to Boston that evening, raced home in the middle of the night, zipping north on the interstate to country roads and finally my home in the forest two thousand miles away. I awoke to the spring I've been patiently observing, my memory of Denver a chimera, contained in a series of evocative memories that I can now place beside this present moment.

I moved through a whole bunch of worlds that weekend and I had lots of opportunities to practice what I've been preaching. I was so grateful to return home so I could sort out what happened, from the familiarity of this place where I have at least a foundation (as limited as it may be) from which to compare things. This is an outpost for biospheric perception and I need its comfort and intimacy. Many of the patterns I perceive occur to me after the experience. The trip to Denver was like a dream and the salient observations are embedded in my memory. They fortify my imagination. I can look at the sky right now from ground level and then recall what it looked like from above the clouds. With both images in mind, from this place and time, I can imagine what today's sky looks like from above. I can compare it to the sky from an hour ago, and have some fun speculating about what it will look like an hour from now. Imagination and memory work together to expand perception. And then when the movement becomes too disconcerting, I return to the present moment.

Thankfully, I have books and observational records to guide me. Photographs, drawings, and recordings freeze time and allow you to compare different spatial and temporal perspectives. Juxtaposing these records allows you to discern patterns in a conceptual way. The more you think about "pattern learning" in any of the domains covered above (interspatial, intertemporal, interspecies, intergenerational), the more equipped you are to recognize them. The first time you go birdwatching you're overwhelmed by the sounds and movement, finding it very difficult to keep up with all of the names and information. With practice, you learn the songs and species, and identification becomes easier. The same is true of biospheric learning. As you study, observe, record, and practice, what was once overwhelming soon becomes familiar. Good teachers know that the learned pattern goes two ways—it gives you a language with which you can venture forward, but it also hides the patterns that only an overwhelmed beginner might detect. Hence the notion "beginner's mind."

Educators must balance all of these learning dynamics. Each student has a different capacity and expertise. We all move through conceptual worlds at different paces. Each of the "biospheric learning patterns" described above corresponds to a predisposition. You can check this out for yourself. Is it easier for you to compare place or time, and what are the circumstances in each case? Here, too, a developmental scheme is help-ful. Perhaps the very young and the very old are best able to master the merging of imagination and memory, while the middle-aged are the best analytical observers. Or perhaps there is no generic lifecycle pattern, and each person proceeds differently according to the learning themes of her life.

For some, book learning is the best template from which to explore vis-ceral perceptual realms. They have to know what they'll be looking at before it's encountered. Otherwise you lose it. Others are visceral learn-ers. They need the body experience before anything they see in a book will make sense. Perhaps it is best to learn a balance of both approaches. Here is a short list of some of the learning paths that have been discussed throughout the book and are ripe for investigation by those who wish to better understand the educational dimensions of biospheric perception:

- Phases of the life cycle (appropriate developmental sequence)
- Balancing imagination, compassion, and analysis
- Place-based paces (how technologies of speed and distance alter perception)
- Interpattern learning (interspatial, interspecies, intertemporal, inter-generational)
- The visceral and the virtual
- The experience of globality

Bringing the biosphere home is a literal message. A place-based orien-tation provides a foundation in natural history and ecology from which a person can venture out and explore. From wherever you are, the bio-sphere is there too! But home can be interpreted in many different ways. Your moral and spiritual home is another good place to start. To inter-pret global environmental change, you have to care about the world and its species. You have to summon your faith in the power of biospheric creation and the human capacity to care for what has been bequeathed.

Abraham Isaac Kook, a Jewish mystic, wrote *The Song of Songs,* a prose poem inspired by the kabbalah. His essential message is that different souls cultivate their love and affiliations in four realms. As you expand your love, these realms expand too, as concentric circles, encompassing broader relationships and considerations, until finally you see that they can't even be distinguished.

There is one who sings the song of his soul, discovering in his soul everything— utter spiritual fulfillment.

There is one who sings the song of his people. Emerging from the private circle of his soul—not expansive enough, not yet tranquil—he strives for fierce heights, clinging to the entire community of Israel in tender love. Together with her he sings her song, feels her anguish, delights in her hopes. He conceives profound insights into her past and her future, deftwardly probing the inwardness of her spirit with the wisdom of love.

Then there is one whose soul expands until it extends beyond the border of Israel, singing the song of humanity. In the glory of the entire human race, in the glory of the human form, his spirit spreads, aspiring to the goal of humankind, envisioning its consummation. From this spring of life, he draws all his deepest reflections, his searching, striving, and vision.

Then there is one who expands even further until he unites with all of existence, with all creatures, with all worlds, singing a song with them all.

There is one who ascends with all these songs in unison—the song of their soul, the song of the nation, the song of humanity, the song of the cosmos—resounding together, blending in harmony, circulating the sap of life, the sound of holy joy.[30]

Biospheric perception is the song of the soul learning to sing earth's music, improvisational melodies and rhythms that you learn to sing in unison with your family, your people, your ecosystem, and your planet. The sound of the world comes in a multiplicity of forms—sonatas, popular tunes, birdsongs, peeper calls, the wind rushing through the trees, thunder, the hum of electricity, ionospheric crackles, the lapping of waves on the shore, and the beating of your heart. There is a great deal of listening to be done. In moments of silent contemplation, the sounds of the biosphere reverberate endlessly, becoming a deafening roar, and you have no choice but to join the band. Life is improvisation and the biosphere is its ever-changing symphony. Our task is to practice its music.

Notes

Chapter 1

1. For more information on biodiversity and species extinction, see Edward O. Wilson, *The Diversity of Life;* Michael J. Jeffries, *Biodiversity and Conservation;* and Andrew P. Dobson, *Conservation and Biodiversity.* For up-to-date assessments of global biodiversity, check out the website of the International Union for Conservation of Nature and Natural Resources (www.redlist.org).

2. For an excellent history of the biosphere concept, see, Jacques Grinevald's introduction to Vladimir I. Vernadsky, *The Biosphere.* Grinevald cites N. Polunin's definition (p. 22) of *biosphere* as the "integrated living and life-supporting system comprising the peripheral element of planet Earth together with its surrounding atmosphere so far down, and up, as any form of life exists naturally." Also helpful for understanding the historical development of the biosphere concept is Paul R. Samson and David Pitt, eds., *The Biosphere and Noosphere Reader.*

3. Global environmental change describes short- and long-term ecological, biogeochemical, physiographic, and anthropogenic changes that occur on a global scale. Of course, the environment is always changing. What's interesting are the trends, cycles, waves, fluctuations, and discontinuities that occur through diverse spatial and temporal realms. Typically, global environmental change refers to the planetary-scale ecological and biogeochemical challenges that confront the human species—climate change, habitat destruction, species extinction, ozone depletion, environmental pollution, and natural resource extraction.

4. Jan DeBlieu's lovely book *Wind* is just such a biospheric meditation.

5. My orientation in *Bringing the Biosphere Home* is to consider the cognitive awareness that is the foundation for perceiving global environmental change. This emphasis should not diminish the importance of understanding how the dynamics of global political economy ultimately influences biodiversity policy.

6. The best single philosophical statement on the cognitive and ecological virtues of place-based environmental learning remains Gary Snyder's essay "The Place, the Region, and the Commons" in The *Practice of the Wild.* For a fine curricular overview, see *Stories in the Land* in the Orion Society's Nature Literacy Series.

7. For an interesting discussion of the political relationship between local and global environmental issues and the importance of social movements and environmental activism,

see Kenneth A. Gould, Allen Schnaiberg, and Adam S. Weinberg, *Local Environmental Struggles: Citizen Activism in the Treadmill of Production*.

8. See Peter Warshall, "Rethinking the Commons" in the fall 1998 issue of *The Whole Earth Review* for a very interesting discussion of the local/global question.

9. Ronnie Lipschutz, *Global Civil Society and Global Environmental Governance*, p. 7.

10. Ibid., p. 22.

11. The issue of population and carrying capacity is also crucial to understanding global environmental change, if beyond the scope of this book. See Joel E. Cohen's encyclopedic *How Many People Can the Earth Support?*

12. An understanding of the political economy of globalization is crucial to interpreting global environmental change. This important theme has been discussed at great length. For a good political critique, start with Jerry Mander and Edward Goldsmith, eds., *The Case against the Global Economy*. Joshua Karliner, in *The Corporate Planet*, cites the specific culpability of various corporations. Mark Hertsgaard's *Earth Odyssey* is an anecdotal trip around the world, with stunning reports of environmental decline.

13. For a clear, accessible, and interesting discussion of biodiversity and the sixth megaextinction, see Niles Eldredge, *Life in the Balance*.

14. The Biodiversity Project's report *Engaging the Public on Biodiversity* describes the gap between the public's concern about environmental quality and its response. The key factors are: "the public's lack of familiarity with many basic ecological principles and processes; the public's lack of connection to nature in daily life; and the deluge of conflicting information that confronts the public." (p. 12)

15. Unquestionably, the mainstream media coverage of global environmental change is spotty, sensationalist, and ephemeral. One wonders how the public would respond if as much coverage was afforded to the loss of biodiversity as is given to the private lives of presidents.

16. A good overview of this confluence is provided by the winter 1997 issue of *The Whole Earth Review*, "The Earth in Crisis: Religion's New Test of Faith." Also see Roger S. Gottlieb's comprehensive anthology, *This Sacred Earth*. Three cheers, too, for the ambitious and comprehensive efforts of Mary Evelyn Tucker and John Grim in organizing a historic series of conferences on world religion and ecology. See the website of the Harvard University Center for the Study of World Religions (www.hds.harvard.edu/cswr/ecology) for essays, links, and forums that pertain to religion and ecology.

17. David Abram's remarkable book, *The Spell of the Sensuous*, inspires the reader to cultivate sensory awareness to perceive the biosphere.

18. Daniel C. Matt, *The Essential Kabbalah*, p. 152.

Chapter 2

1. See the *New York Times* of January 1, 2000. The vertical alignment of juxtaposed midnights appears on pp. A2 and A3. "Future Threatens a Place without Calendars" is written by Rachel L. Swarns and appears on p. 5.

2. Frank White's book *The Overview Effect* is a comprehensive discussion and series of interviews depicting how astronauts' views of the earth dramatically changed after their space voyages.

3. Anderson, *Imagined Communities*, p. 33.

4. Ibid., 35. The paraphrases from Anderson, "imagined linkage," "calendrical coincidence," "Steady onward clocking of homogeneous time," and "one day best seller," may be found on pp. 33 to 35.

5. Marsh, *Man and Nature*, p. 3.

6. Bowler, *The Norton History of the Environmental Sciences*, p. 6.

7. Ravenstein's work is profiled by Joel E. Cohen in *How Many People Can the Earth Support?* Cohen's superb and comprehensive review of carrying capacity includes a detailed historical interpretation of the idea.

8. This summary is adapted from the foreword (written by thirteen scientists) to the English-language edition of Vladimir I. Vernadsky, *The Biosphere*.

9. Sagan, *Biospheres*, p. 41.

10. See Donella Meadows, *Beyond the Limits*, for the most up-to-date iteration of this approach.

11. Since 1984, the Worldwatch Institute has published its excellent *State of the World* yearbooks. Since 1992, they have published *Vital Signs: The Environmental Trends That Are Shaping Our Future*. Reviewing these from the beginning provides a fine recent history of the emergence of various global change issues. Appropriately, you can now also obtain an annual Worldwatch database disk (www.worldwatch.org).

12. This landmark volume, published by W.H. Freeman in 1970, contains G. Evelyn Hutchinson's classic article, "The Biosphere." The anthology is organized from a biogeochemical perspective with articles on cycles of energy, water, carbon, hydrogen, oxygen, and nitrogen.

13. For a class I teach on global environmental change, as a first assignment I ask students to choose a global change text to profile, indicating how the author organizes the material. This reveals that global environmental change is interpreted differently depending on the interests and backgrounds of the author. One would expect this since the material is so vast. Nevertheless, all of these texts share a similar problem—how to best conceptually organize this vast array of material. A survey of these texts is an interesting way to see how the discourse of global environmental change emerges. In particular see the excellent, but very differently organized texts by Mannion, Goudie, Simmons, and Huggett.

14. See David Takacs, *The Idea of Biodiversity*, for a discussion of how the concept of biodiversity is "constructed."

15. Two good introductory texts on global climate change, which also give you a good sense of the discovery of global warming, are John Houghton, *Global Warming*, and Thomas E. Graedel and Paul J. Crutzen, *Atmosphere, Climate and Change*.

16. The best single volume covering the history and ramifications of Gaia is Stephen H. Schneider and Penelope J. Boston, *Scientists on Gaia*.

17. For a good, succinct review of global change theory until 1990, see Robert W. Kates, B. L. Turner II, and William C. Clark, "The Great Transformation" in *The Earth as Transformed by Human Action*. Also see "Problems of Theory and Method" in the National Research Council's *Global Environmental Change*.

18. Volk is particularly interested in the use of metaphors in science. See his earlier work, *Metapatterns*, as well as *Gaia's Body*.

19. Connie Barlow's *Green Space, Green Time* covers the narrative richness of themes such as the biosphere, biodiversity, and biogeochemical cycling.

20. Using the material habits of ordinary life as a means to teach environmental studies is highly effective. I cover this approach at some length in *Ecological Identity*. Also see John C. Ryan and Alan Thein Durning, *Stuff: The Secret Lives of Everyday Things*.

21. Some people learn about global environmental change by being directly exposed to its most visible effects, let's say Hurricane Mitch. Yet without a scientific context, it's very difficult to link such a disaster to broader biopsheric patterns. Rather, one is left with all of the pain and turmoil that surrounds a natural catastrophe—how to recuperate, where to go from here, how to return to some sense of normalcy in the face of destruction. This is hardly a way to contemplate the biosphere. Nevertheless, disasters such as these provide environmental educators with an opportunity to provoke questions regarding biospheric processes. What are the environmental causes and consequences of events such as these? Surely a master narrative of the biosphere can be evoked.

Chapter 3

1. The environmental news service provides daily, online coverage of environmental issues.

2. Cited in Herbert Fingarette, *The Self in Transformation*, p. 73.

3. In chapter 2, "Anxiety and Disintegration," Fingarette addresses the difficulties in distinguishing between "pathological" or "neurotic" anxiety and "ontological" anxiety. In this chapter I am referring to the philosophical aspects of meaning and being.

4. Robert J. Lifton discusses the psychoanalytic impact of the imagery of annihilation. For a succinct statement about this, see the section "The World is Ending," pp. 21–24, in *The Protean Self*.

5. Throughout the book I use the term "creation" in a scientific sense, referring to the great variety of life forms that emerge from ecological, evolutionary, and biogeochemical processes. Nevertheless, the biospheric implications of creation are of great spiritual as well as scientific interest.

6. The "nearest order of magnitude" quote is from p. 140 of *The Diversity of Life*. "The Unexplored Biosphere" is the title of chapter 8.

7. Dobson, *Conservation and Biodiversity*, p. 30.

8. Dobson lays out these figures on p. 24. Of particular interest is an illustration on page 25 which depicts the size of organisms according to their relative contribution to total biodiversity. It is sobering to see a giant beetle next to a tiny bear.

9. Tudge, *The Variety of Life*, p. 9. See chapter 1, "So Many Goodly Creatures," for a compelling discussion about the importance of taxonomy in appreciating biodiversity.

10. For a comprehensive discussion of the status of biodiversity in the United States, see Stein, Kutner, and Adams, *Precious Heritage*. Look up any section of North America and find out who your neighbors are and which are endangered or extinct, as well as what's being done to protect them.

11. Wilson, *The Diversity of Life*, p. 142.

12. Ibid.

13. Here's Wilson's definition of *biodiversity* from *The Diversity of Life*, p. 393: "The variety of organisms considered at all levels, from genetic variants belonging to the same species through arrays of species to arrays of genera, families, and still higher taxonomic levels; including the variety of ecosystems, which comprise both the communities of organisms within particular habitats and the physical conditions under which they live."

14. Wilson, *The Diversity of Life*, p. 151.

15. Fortey, *Life*, p. 209.

16. Huggett, *Environmental Change*, p. 300.

17. For more on the specifics of the various megaextinctions, see Steven M. Stanley, *Extinction*.

18. Wilson, *The Diversity of Life*, p. 31.

19. Dobson, p. 71. See chapter Three "The Mathematics of Extinction" in *Conservation and Biodiversity* to see how these figures are derived.

20. Ibid., p. 61.

21. Wilson, *The Diversity of Life*, p. 255

22. For two interesting accounts that reveal the emotional loss experienced by scientists who witness species extinction, see Beverly Peterson Stearns and Stephen C. Stearns, *Watching, from the Edge of Extinction*, and Diane Ackerman, *The Rarest of the Rare*.

23. Eldredge, *Life in the Balance*, p. 66.

24. For two good summaries of these arguments, see Wilson, *The Diversity of Life*, chapter 14, "Resolution," and Niles Eldredge, *Life in the Balance*, chapter 5, "Biodiversity—A Threatened National Treasure." For a philosopher's approach, see Holmes Rolston, *Conserving Natural Value*.

25. Lawrence Thornton, in *Imagining Argentina*, raises two issues pertinent to the theme of this chapter—how people keep alive in the midst of seemingly overwhelming circumstances and the role of imagination in fostering hope.

26. See Pyle, *Chasing Monarchs*, for a compelling account of this journey.

27. Bill McKibben was one of the writers who visited the sanctuary.

28. Heschel, *Man Is Not Alone*, p. 13.

29. Heschel, *God in Search of Man*, p. 36.

30. Daston and Park, *Wonder and Order of Nature*, p. 14.

31. Ibid.

32. Heschel, *God in Search of Man*, p. 51.

33. Heschel, *Man Is Not Alone*, p. 12.

34. Heschel, *God in Search of Man*, p. 49.

35. Ibid.

36. Ibid., p. 85.

37. Ibid., p. 43.

38. See Robert Michael Pyle's account, "When Things Go Wrong," in *Orion Afield*, summer 2000.

39. The conservation commission realized that had it spent more time working with landowners prior to the vote, the outcome might have been different. Now its strategy is to implement similar measures, but to do so at a town level.

40. Heschel, *The Prophets*. Vol. 1, p. 12.

41. Ibid., p. 21.

42. I am making the linkage between biospheric advocates and biblical prophets to consider some of the issues that emerge when citizens and educators hold the public to task for their failure to recognize various threats to the biosphere. Of course, many people are inspired both by the biosphere and by some divine source, or have a theological vision that integrates them.

43. Heschel, *The Prophets*, p. 12.

44. Ibid., p. 16.

45. Ibid., p. xii.

46. For more on Heschel, see the two fine works by Edward K. Kaplan. *Holiness in Words* is an illuminating interpretation of Heschel's work. Also see the biography of Heschel by Kaplan and Samuel H. Dresner, *Abraham Joshua Heschel: Prophetic Witness*.

47. Heschel, *Man Is Not Alone*, p. 185.

48. Ibid., p. 69.

49. Ibid.

50. Sanders, *Hunting for Hope*, p. 9.

51. Ibid.

52. Ibid., p. 185.

53. Ibid., p. 22.

54. Ibid., p. 23.

55. The Maimonides quote is taken from Irving Greenberg, *The Jewish Way*, p. 190.

56. Sanders, p. 57.

Chapter 4

1. For more on the relationship between perception and environmental learning, see Laura Sewall, *Sight and Sensibility*.

2. Abram, p. ix.

3. Thomashow, *Ecological Identity*. See the section "The Sense of Place Map" in chapter 6, pp. 192–199.

4. See Michael McGinnis, ed., *Bioregionalism*, for a thorough review of the relationship between sense of place, political economy, and bioregionalism.

5. Sobel, *Mapmaking with Children*, p. 21.

6. Jim Dodge, "Living by Life: Some Bioregional Theory and Practice," *CoEvolution Quarterly*, no. 32, pp. 6–12, 1981.

7. Gary Paul Nabhan and Sara St. Antoine, "The Loss of Floral and Faunal Story: The Extinction Experience," in Stephen R. Kellert and Edward O. Wilson, *The Biophilia Hypothesis*.

8. Robert Michael Pyle, p. 145

9. Kahn, p. xvii.

10. Nabhan and Trimble, p. 97.

11. John Elder, *Stories in the Land*, pp. 14–15.

12. Wilson, *The Diversity of Life*, p. 4.

13. Meeker, p. 2.

14. Bateson, p. 9.

15. An exemplary approach is that of Clare Walker Leslie and Charles E. Roth, *Nature Journaling*. This is one of the best combinations of natural history and art instruction I've seen. It is filled with methods and approaches that are relevant to place-based perceptual ecology, and absolutely reflect the spirit of this chapter.

16. Thoreau, *Walden*, p. 343.

17. Ackerman, *A Natural History of the Senses*, p. 5.

18. Gary Snyder suggests in *The Practice of the Wild*, p. 153: "The truly experienced person, the refined person, delights in the ordinary."

19. Schafer, p. 24.

20. Steven Feld, "Waterfalls of Song: An Acoustemology of Place Resounding in Bosavi, Papua New Guinea," in Steven Feld and Keith H. Basso, eds., *Senses of Place*, p. 98.

21. Thoreau, "Walking," p. 598.

22. Mitchell, p. 25.

23. Ibid., p. 33.

24. Ibid., p. 1.

25. Ibid., p. 12.

26. The practice of environmental history melds landscape, ecology, and community. See the classic works by William Cronon, *Changes in the Land*, and Carolyn Merchant, *Ecological Revolutions*.

27. See Jacob von Uexkull, "A Stroll Through the World of Animals and Men: A Picture Book of Invisible Worlds," in Claire H. Schiller, ed., *Instinctive Behavior*. This is a delightful essay which deserves a much wider audience. For more on sensory perception and animals, see Howard C. Hughes, *Sensory Exotica*, and Donald Griffin, *Animal Minds*.

28. Allen and Hoekstra, p. 173.

29. Rothenberg, p. xvi.

30. Allen and Hoekstra, p. 9.

31. In moving through this conceptual sequence, I'm using a "methodology" derived from scientific ecology and natural history. This doesn't exclude other ways of learning these patterns. A shamanic approach shifts perceptual scale by entering the "spirit world" of a species. For an interesting fictional representation of this approach, see the short story by Hermann Hesse, "The Rainmaker," which is appended to *The Glass Bead Game*. Drought is apprehended viscerally through direct contact with the spirit of the landscape. Are such ways of knowing merely another way of communicating intimate awareness of natural history?

32. See Abram, chapter 1, "The Ecology of Magic."

33. Elder, *Reading the Mountains of Home*, p. 33.

34. Krafel, p. 52.

35. Ibid., p. 62.

36. John C. Kricher and Gordon Morrison, *Eastern Forests*.

Chapter 5

1. Vernadsky, p. 61.

2. Bachelard, *The Poetics of Space*, p. 9.

3. Consider the elegant system of the ancient *I Ching*, the *Book of Changes*, derived from biospheric archetypes (heaven, lake, fire, thunder, wind, water, mountain, earth). The movement of the archetypes provides a code for studying nature, human behavior, laws and government, ethics and virtue. Endless transformations and repeated cycles emerge from the interplay of these archetypes (trigrams). Water over Mountain. Heaven over Earth. You learn to read the conditions of any situation based on the flow of elemental movement. Patterns of the biosphere, patterns of mind. Observe the biosphere and learn how to interpret the world at multiple scales. Learn the path of virtue in harmony with the elements.

4. Vernadsky, 60.

5. See Philip W. Connkling, ed., *From Cape Cod to the Bay of Fundy: An Environmental Atlas of the Gulf of Maine* for further information on all of these themes.

6. Vernadsky, p. 72.

7. For more on Vernadsky's life and times, see Kendall E. Bailes, *Science and Russian Culture in an Age of Revolutions*.

8. Fortey, 4.

9. Ibid., p. 10.

10. Ibid., p. 119.

11. Ibid., p. 52.

12. Ibid., p. 149.

13. Ibid., p. 167.

14. Ibid., p. 121.

15. See Fortey's more recent book, *Trilobite! Eyewitness to Evolution.*

16. Margulis, p. 69.

17. Ibid., p. 70.

18. Ibid.

19. See Margulis, *Symbiotic Planet,* for a readable, semiautobiographical discussion of her work and a sense of her legacy for biosphere science. It covers two of her major theoretical contributions, SET (serial endosymbiosis theory), and the Gaia Hypothesis.

20. Volk, *Gaia's Body,* p. 17.

21. Ibid., p. 14.

22. Ibid., p. 15.

23. One of the best approaches to learning the geological time scale can be found in Nigel Calder, *Timescale.* Unfortunately, it's out of print. Also see Stephen Jay Gould, ed., *The Book of Life.* John McPhee, in *Annals of the Former World* (pp. 69–99), presents this material as lucidly and efficiently as I've seen. Another rich book is Stephen Drury, *Stepping Stones.* To understand the time scale as applied to the history of North America, see David Rockwell, *The Nature of North America.*

24. For a full elaboration of the five kingdom perspective see Margulis and Schwartz, *Five Kingdoms,* and Tudge, *The Variety of Life.*

25. Volk's *Gaia's Body* is a fine introduction to biogeochemical cycles from a Gaian perspective. Also, see the beautifully illustrated and readable book by Vaclav Smil, *Cycles of Life.*

26. See Paul F. Berliner, *Thinking in Jazz,* for a comprehensive study of how jazz musicians learn to balance structure and improvisation.

27. Stephen Jay Gould, 20.

28. Ibid., p. 13.

29. Another terrific out-of-print book. A fine, hands-on portrait of the five kingdoms that pays great attention to the juxtaposition of scale.

30. Bachelard, *Air and Dreams: An Essay on the Imagination of Movement,* p. 1. See also Bachelard's *Water and Dreams: An Essay on the Imagination of Matter* for interesting literary approaches to biospheric phenomena.

31. Volk, *Gaia's Body,* p. 190.

32. Ibid.

33. Ibid., p. 250.

34. Margulis and Sagan, *What is Life?,* p. 194.

35. Fortey, p. 322.

36. Ibid.

37. For a comprehensive discussion of the duration of environmental change, see Patricia F. McDowell, Thompson Webb III, and Patrick J. Bartlein, "Long-Term Environmental Change," in B.L. Turner et al., eds., *The Earth as Transformed by Human Action*. Also see chapter 1, "Introducing Environmental Change," in Huggett's *Environmental Change*.

38. Huggett, p. 6.

39. Allen and Hoekstra, p. 20.

40. Ibid., p. 299.

41. Margulis, Schwartz, and Dolan, *Diversity of Life*, p. 36.

42. Margulis and Schwartz, p. 72.

43. Smil provides a thorough and interesting introduction to the biogeochemical cycles, with a special emphasis on the role of anthropogenic activity in altering those cycles.

44. Margulis and Sagan, *What is Life?*, p. 175.

45. Margulis, Schwartz, and Dolan, p. vii.

46. Other compelling approaches to "tracking" ancestry and lineage are the paleontological (finding fossil tracks) and the paleoecological (reconstructing past environments). See Martin Lockley's *The Eternal Trail* for a philosophical approach to finding fossil footprints. See Neil Roberts, *The Holocene: An Enviromental History* for a hands-on introduction to some of the techniques paleoecologists and geologists use in tracking the recent past.

Chapter 6

1. Simply put, speed refers to how much time it takes to do something. The fastest computers are determined by the megahertz of the chip, indicating greater processing power. With information, speed is measured via bits per second. So when you visit a website using a browser, running along the bottom of the "window" is a measurement of bits per second, keeping tabs on how quickly the information is being conveyed. With transportation, speed refers to how much time it takes to cover a set distance. In reference to moving humans from place to place, we refer to miles per hour. One of the remarkable qualities of modern transportation is how fast you move between places. The faster you move, the more still your surroundings appear. You can travel at high speeds in a jet, with a book in hand or dinner in your lap, while you gaze down thirty thousand feet at a landscape that appears entirely still.

2. There are many "reflective" commentaries on Internet technologies and the use of computers. See James Gleick, *Faster* for a discussion of speed and expectation, Mitchell Stephens, *The Rise of the Image, the Fall of the Word*, for an interesting view on the meaning of images replacing reading, and see Sven Birkerts, *The Gutenberg Elegies*, for a critique of the sensibility of computers. My point is that most people embrace new technologies without giving serious thought to their perceptual impact. For a philosopher's view of information, see Albert Borgmann, *Holding on to Reality*. To my mind, the most applicable and interesting philosophy of technology, especially regarding how the use of technology changes one's perceptions of nature, remains David Rothenberg's *Hand's End*.

3. Lopez, *About This Life*, p. 93.

4. Perlman and Adelson, p. 75.

5. Ibid., p. 78.

6. Ibid., p. 81.

7. Ibid., p. 83.

8. National Geographic, *Satellite Atlas of the World*, p. 24. Also see Apt, Helfert, and Wilkinson, *Orbit*, and Strain and Engle, *Looking at Earth*.

9. Ibid.

10. Mark Monmonier, in *Air Apparent*, notes the importance of computers and mapping for understanding global climate change:

Our best defense against catastrophic climate change rests on a four-fold strategy of monitoring, modeling, archiving and visualizing. In addition to revealing less subtle environmental hazards like volcanic eruptions and the ozone hole, networks of satellite and surface sensors provide a foundation for numerical models that test hypotheses about ocean-atmosphere interactions as well as predict the consequences of remediation treaties or new dangers. Archiving provides an information base essential for validating long-range models and assessing the significance of newly discovered anomalies. Visualization supports the other three strategies by helping researchers and policy makers cope with an increasingly overwhelming variety of intriguing, highly complex maps and by encouraging the data to speak for themselves. (pp. 211–212)

11. Forman, p. 35.

12. Hall, p. 4.

13. Ibid., p. 12.

14. Ibid., p. 21.

15. Forman, p. 34.

16. Wilson, *Consilience*, p. 269.

17. Citation taken from the Cornell Laboratory of Ornithology website (http://birds.cornell.edu).

18. Lopez, p. 143.

19. Ryan, p. 113.

20. Ibid.

21. Ibid., p. 114.

22. Ibid., p. 115.

23. Rothenberg, p. 14.

24. Ibid., p. 215.

25. Ibid., p. xvii.

26. Gary Snyder, *Turtle Island*, p. 100.

27. See the essay of that title in Paul Shepard, *Traces of an Omnivore*.

28. A study of the political economy of computer technology reminds us of the unconscionable rate of planned obsolescence. The environmental consequences of this process should not be overlooked.

Chapter 7

1. Clifford, p. 26.

2. See Pielou, *After the Ice Age*, for an excellent discussion of the migration of North American flora and fauna since the retreat of the glaciers.

3. For more on the sociology of American migration patterns, see Leach, *Country of Exiles.*

4. This quote comes from "New Immigrant Tide: Shuttle Between Worlds in The New York Times, July 19, 1998. The other citations are from the three-part sequence of articles, also including from "Wedding Vows Bind Old Work and New" (July 20) and "A Mexican Town That Transcends All Borders" (July 21).

5. Sowell, p. 1.

6. R. Cole Harris, ed., *Historical Atlas of Canada*, plate 14.

7. These figures are taken from the fact sheet on migration at the "one world" website (www.oneworld.org/ni/issue305/facts.html). Also see the website of the International Organization for Migration (www.iom.ch/).

8. Weidensaul, p. 8.

9. Waterman, p. 31. All of the examples in this section are from Waterman. He defines migration (page 7) as "periodic cyclic movement from one part of an animal's habitat to another and back."

10. For more on the controversy surrounding the community concept in ecology, see Allen and Hoekstra, chapter 4, "The Community Criterion," in *Toward a Unified Ecology.*

11. All of the Huggett citations come from his sections on community in *Environmental Change*, pp. 269–275.

12. Huggett, p. 288.

13. Ibid.

14. Ibid.

15. Chaliand and Rageau, pp. xiv, xv.

16. Clifford, p. 255.

17. Ibid.

18. Cohen, p. 26.

19. Clifford, p. 257.

20. Ibid.

21. Ibid.

22. Some additional approaches to diaspora: a good evolutionary and genetic perspective can be found in Cavalli-Sforza, Luca, and Cavalli-Sforza, *The Great Human Diasporas*; Fritz reveals the dark side of forced migration in *Lost on Earth*; in *Tribes*, Kotkin advocates diaspora as a means to gain economic advantage, Dorfman, *Heading South, Looking North* discusses diaspora and political identity, and Glendinning, *Off the Map*, bemoans the imperial consequences of misguided mobility.

23. Heschel, *The Sabbath*, p. 8.

24. Ibid., p. 10.

25. Abram, p. 140.

26. Mann, p. 3.

27. Nabhan, *Cultures of Habitat*, p. 13.

28. Margulis and Sagan, *What is Life?*, p. 173.

29. See chapter 6, "La Selva Maia," in Weidensaul, pp. 129–152.

30. Robin Cohen, p. 196.

31. Cohen's gardening guide to diasporas is summarized in a table on p. 178. All of the subsequent citations from Cohen are from this table.

32. See Meinig, pp. 65–76.

33. Kaplan's journalistic, impressionistic approach is a powerful, if opinionated, view on the inseparability of borders.

34. Weidensaul, p. 107.

35. Nabhan, *Cultures of Habitat*, p. 37.

36. Ibid.

37. Pinker describes the relationship between linguistic diversity and biodiversity. For an excellent book-length treatment of this topic, see Nettle and Romaine, *Vanishing Voices*.

38. See Ostrom's classic work, *Governing the Commons*.

39. For more on the pollination crisis, see Buchmann and Nabhan, *The Forgotten Pollinators*.

Chapter 8

1. Gardner, *Intelligence Reframed*.

2. Wilson, "Biophilia and the Conservation Ethic," p. 31, in Kellert and Wilson.

3. See the excellent, comprehensive landscape ecology text, *Land Mosaics*, by Richard T. T. Forman for a full elaboration of these concepts.

4. For more details on the theory of island biogeography, see Wilson, *The Diversity of Life*, pp. 200–227. Also see Quammen, *The Song of the Dodo*.

5. Ball, p. 111.

6. Morrison and Morrison.

7. See Alexander, Ishikawa, and Silverstein. They use the patterns of spatial relationships as a means to develop an ecological view of regional planning.

8. Nelson, *Make Prayers to the Raven*, p. 16.

9. Ibid., p. 17.

10. Shepard, *Traces of an Omnivore*, p. 56.

11. Ibid., p. 68.

12. Ibid., p. 70.

13. See Margulis and Sagan. *Gaia to Microcosm*. Vol. 1. Dubuque, Iowa: Kendall/Hunt Publishing Co., 1996. Videocassette.

14. Uexkull, p. 5.

15. "Empathy with lives that are alien to our own is the human impulse that gives rise to vernacular practices that celebrate and regulate our links to other species." House, p. 99.

16. I know myself as a predator, know the hunter inside me, know the communion of meat and blood that shapes my body from those of deer." Nelson, *Heart and Blood*, p. 7.

17. Sobel, *Beyond Ecophobia*, pp. 14–15.

18. Fraser, p. 67.

19. See chapter 2, "Reconstructing Holocene Environments," in Roberts, *The Holocene*.

20. "Fossil footprints, while being themselves tangible, are like scripts that tell of the less tangible dimensions of existence of extinct animals. They tell of behavior and give us subtle insights into the spirit of animals and how they interacted with contemporary species and the environments that they called home." Lockley, p. 9.

21. Adam, p. 54

22. Ibid., p. 11

23. Heschel, *Who Is Man?*, p. 114.

24. Heschel, *Man Is Not Alone*, p. 194.

25. Heschel, *God in Search of Man*, p. 353.

26. See Heschel's chapter "The Meaning of Observance," pp. 348–366, in *God in Search of Man* for a discussion of mitzvoth.

27. Ibid., p. 357.

28. Green, pp. 84–86.

29. Thich Nhat Hanh, p. 13.

30. Matt, p. 154.

Bibliography

Abram, David. *The Spell of the Sensuous: Perception and Language in a More-Than-Human World*. New York: Random House, 1996.

Ackerman, Diane. *A Natural History of the Senses*. New York: Random House, 1990.

Ackerman, Diane. *The Rarest of the Rare*. New York: Random House, 1997.

Adam, Barbara. *Timescapes of Modernity*. New York: Routledge, 1998.

Alexander, Christopher, Sara Ishikawa, and Murray Silverstein. *A Pattern Language*. New York: Oxford University Press, 1977.

Allen, Timothy F. H., and Thomas W. Hoekstra. *Toward a Unified Ecology*. New York: Columbia University Press, 1992.

Anderson, Benedict. *Imagined Communities*. New York: Verso, 1983.

Apt, Jay, Michael Helfert, and Justin Wilkinson. *Orbit: NASA Astronauts Photograph the Earth*. Washington, DC: National Geographic Society, 1996.

Bachelard, Gaston. *Water and Dreams: An Essay on the Imagination of Matter*. Dallas: The Dallas Institute of Humanities and Culture, 1983.

Bachelard, Gaston. *Air and Dreams: An Essay on the Imagination of Movement*. Dallas: The Dallas Institute of Humanities and Culture, 1988.

Bachelard, Gaston. *The Poetics of Space*. Boston: Beacon Press, 1994.

Bailes, Kendall E. *Science and Russian Culture in an Age of Revolutions: V. I. Vernadsky and His Scientific School, 1863–1945*. Bloomington: Indiana University Press, 1990.

Bailey, Robert. *Description of the Ecoregions of the United States*. Washington, DC: USDA Forest Service, 1995.

Ball, Philip. *The Self-Made Tapestry: Pattern Formation in Nature*. New York: Oxford University Press, 1999.

Barlow, Connie. *Green Space, Green Time: The Way of Science*. New York: Springer-Verlag, 1997.

Bateson, Mary Catherine. *Peripheral Visions: Learning along the Way*. New York: HarperCollins, 1994.

Berliner, Paul F. *Thinking in Jazz: The Infinite Art of Improvisation*. Chicago: University of Chicago Press, 1994.

Biodiversity Project, *Engaging the Public on Biodiversity*. Madison, WI: The Biodiversity Project, 1998.

Birkerts, Sven. *The Gutenberg Elegies: The Fate of Reading in an Electronic Age*. Boston: Faber & Faber, 1993.

Borgmann, Albert. *Holding on to Reality: The Nature of Information at the Turn of the Millennium*. Chicago: University of Chicago Press, 1999.

Bowler, Peter J. *The Norton History of the Environmental Sciences*. New York: W. W. Norton, 1993.

Brand, Stewart. *The Clock of the Long Now*. New York: Basic Books, 1999.

Brown, Harrison. *The Challenge of Man's Future*. New York: Viking, 1954.

Brown, Lester R., Michael Renner, and Brian Halweil. *Vital Signs 1999*. New York: W. W. Norton, 1999.

Buchmann, Stephen L., and Gary Paul Nabhan. *The Forgotten Pollinators*. Washington, DC: Island Press, 1996.

Calder, Nigel. *Timescale: An Atlas of the Fourth Dimension*. New York: Penguin, 1983.

Cavall-Sforza, Luigi Luca, and Francesco Cavalli-Sforza. *The Great Human Diasporas: The History of Diversity and Evolution*. Reading, MA: Addison-Wesley, 1995.

Chaliand, Gerard, and Jean-Pierre Ragueu. *The Penguin Atlas of Diasporas*. New York: Viking, 1995.

Chawla, Louise. *In the First Country of Places: Nature, Poetry, and Childhood Memory*. Albany: State University of New York Press, 1994.

Clifford, James. *Routes: Travel and Translation in the Late Twentieth Century*. Cambridge, MA: Harvard University Press, 1997.

Cobb, Edith. *The Ecology of Imagination in Childhood*. Dallas: Spring Publications, 1993.

Cohen, Joel E. *How Many People Can the Earth Support?* New York: W. W. Norton, 1995.

Cohen, Robin. *Global Diasporas*. Seattle: University of Washington Press, 1997.

Conkling, Philip, W., ed. *From Cape Cod to the Bay of Fundy: An Environmental Atlas of the Gulf of Maine*. Cambridge, MA: The MIT Press, 1995.

Cronon, William. *Changes in the Land: Indians, Colonists, and the Ecology of New England*. New York: Hill & Wang, 1983.

Daston, Lorraine, and Katharine Park, *Wonders and the Order of Nature*. New York: Zone Books, 1998.

DeBlieu, Jan. *Wind: How the Flow of Air Has Shaped Life, Myth and the Land*. Boston: Houghton Mifflin, 1998.

Dobson, Andrew P. *Conservation and Biodiversity*. New York: W. H. Freeman, 1998.

Dorfman, Ariel. *Heading South, Looking North*. New York: Penguin, 1998.

Drury, Stephen. *Stepping Stones: The Making of Our Home World.* New York: Oxford University Press, 1999.

Elder, John. *Reading the Mountains of Home.* Cambridge, MA: Harvard University Press, 1998.

Elder, John, ed. *Stories in the Land. A Place-Based Environmental Education Anthology.* Great Barrington, MA: The Orion Society, 1998.

Eldredge, Niles. *Life in the Balance: Humanity and the Biodiversity Crisis.* Princeton, NJ: Princeton University Press, 1998.

Feld, Steven, and Keith H. Basso, eds., *Senses of Place.* Santa Fe: School of American Research Press, 1996.

Fingarette, Herbert. *The Self in Transformation: Psychoanalysis, Philosophy and the Life of the Spirit.* New York: Harper & Row, 1965.

Forman, Richard T. T. *Land Mosaics: The Ecology of Landscape and Regions.* New York: Cambridge University Press, 1995.

Fortey, Richard. *Life: A Natural History of the First Four Billion Years of Life on Earth.* New York: Alfred A. Knopf, 1998.

Fortey, Richard. *Trilobite: Eyewitness to Evolution.* New York: Alfred A. Knopf, 2000.

Fraser, J. T. *Of Time, Passion, and Knowledge.* New York: George Braziller, 1975.

Fritz, Mark. *Lost on Earth: Nomads of the New World.* Boston: Little, Brown, 1999.

Gardner, Howard. *Intelligence Reframed: Multiple Intelligences for the Twenty-First Century.* New York: Basic Books, 1999.

Gleick, James. *Faster: The Acceleration of Just about Everything.* New York: Random House, 1999.

Glendinning, Chellis. *Off the Map: An Expedition into Imperialism, the Global Economy, and Other Earthly Whereabouts.* Boston: Shambhala, 1999.

Gottlieb, Roger S., ed. *This Sacred Earth: Religion, Nature, Environment.* New York: Routledge, 1995.

Goudie, Andrew. *The Human Impact on the Natural Environment.* Oxford: Blackwell, 1995.

Gould, Kenneth A., Allen Schnaiberg, and Adam S. Weinberg, *Local Environmental Struggles: Citizen Activism in the Treadmill of Production.* New York: Cambridge University Press, 1996.

Gould, Stephen Jay, ed. *The Book of Life: An Illustrated History of the Evolution of Life on Earth.* New York: W. W. Norton, 1993.

Graedel, Thomas E., and Paul J. Crutzen, *Atmosphere, Climate and Change.* New York: W. H. Freeman, 1997.

Green, Arthur. *Seek My Face, Speak My Name: A Contermporary Jewish Theology.* Northvale, NJ: Jason Aronson, 1992.

Greenberg, Irving. *The Jewish Way: Living the Holidays.* New York: Simon & Schuster, 1988.

Griffin, Donald. *Animal Minds.* Chicago: University of Chicago Press, 1992.

Hall, Stephen S. *Mapping the Next Millennium*. New York: Random House, 1992.

Harris, R. Cole, ed. *Historical Atlas of Canada*. Toronto: University of Toronto Press, 1987.

Harvey, David. *Justice, Nature and the Geography of Difference*. Cambridge, MA: Blackwell, 1996.

Hertsgaard, Mark. *Earth Odyssey: Around the World in Search of Our Environmental Future*. New York: Random House, 1996.

Heschel, Abraham Joshua. *Who Is Man?* Stanford, CA: Stanford University Press, 1965.

Heschel, Abraham Joshua. *The Prophets*. Vol. 1. New York: Harper & Row, 1969.

Heschel, Abraham Joshua. *Man Is Not Alone*. New York: Farrar, Straus & Giroux, 1976.

Heschel, Abraham Joshua. *God in Search of Man*. New York: Farrar, Straus & Giroux. 1976.

Heschel, Abraham Joshua. *The Sabbath*. New York: Farrar, Straus & Giroux, 1977.

Houghton, John. *Global Warming*. New York: Cambridge University Press, 1997.

House, Freeman. *Totem Salmon: Life Lessons from Another Species*. Boston: Beacon Press, 1999.

Huggett, Richard John. *Environmental Change: The Evolving Ecosphere*. New York: Routledge, 1997.

Hughes, Howard C. *Sensory Exotica: A World Beyond Human Experience*. Cambridge, MA: The MIT Press, 1999.

Jeffries, Michael J. *Biodiversity and Conservation*. New York: Routledge, 1997.

Kahn, Peter H. *The Human Relationship with Nature: Development and Culture*. Cambridge, MA: The MIT Press, 1999.

Kaplan, Edward K. *Holiness in Words: Abraham Joshua Heschel's Poetics of Piety*. Albany: State University of New York Press, 1996.

Kaplan, Edward K., and Samuel H. Dresner. *Abraham Joshua Heschel: Prophetic Witness*. New Haven, CT: Yale University Press, 1998.

Kaplan, Robert D. *An Empire Wilderness: Travels into America's Future*. New York: Random House, 1998.

Karliner, Joshua. *The Corporate Planet*. San Francisco: Sierra Club Books, 1997.

Kates, Robert W., B. L. Turner II, and William C. Clark. *The Earth as Transformed by Human Action*. New York: Cambridge University Press, 1990.

Kellert, Stephen R., and Edward O. Wilson, *The Biophilia Hypothesis*. Washington, DC: Island Press, 1993.

Kotkin, Joel. *Tribes: How Race, Religion, and Identity Determine Success in the New Global Economy*. New York: Random House, 1992.

Krafel, Paul. *Seeing Nature: Deliberate Encounters with the Visible World*. White River Junction, VT: Chelsea Green, 2000.

Kricher, John C., and Gordon Morrison. *Eastern Forests*. Boston: Houghton Mifflin, 1988.

Leach, William. *Country of Exiles:The Destruction of Place in American Life*. New York: Random House, 1999.

Leslie, Clare Walker, and Charles E. Roth. *Nature Journaling*. Pownal, VT: Storey Books, 1998.

Lifton, Robert Jay. *The Protean Self: Human Resilience in an Age of Fragmentation*. New York: Basic Books, 1993.

Lipschutz, Ronnie. *Global Civil Society and Global Environmental Governance*. Albany: State University of New York Press, 1996.

Lockley, Martin. *The Eternal Trail: A Tracker Looks at Evolution*. Reading, MA: Perseus Books, 1998.

Lopez, Barry. *About This Life: Journeys on the Threshold of Memory*. New York: Random House, 1998.

Mander, Jerry, and Edward Goldsmith, eds. *The Case against the Global Economy*. San Francisco: Sierra Club Books, 1996.

Mann, Thomas. *Joseph and His Brothers*. New York: Alfred A. Knopf, 1963.

Mannion, A. M. *Global Environmental Change: A Natural and Cultural Environmental History*. New York: Longman, 1997.

Margulis, Lynn. *Symbiotic Planet: A New View of Evolution*. New York: Basic Books, 1998.

Margulis, Lynn, and Dorion Sagan. *The Microcosmos Coloring Book*. New York: Harcourt Brace Jovanovich, 1988.

Margulis, Lynn, and Dorion Sagan. *What is Life?* New York: Simon & Schuster, 1995.

Margulis, Lynn, and Karlene V. Schwartz. *Five Kingdoms: An Illustrated Guide to the Phyla of Life on Earth*. New York: W. H. Freeman, 1997.

Margulis, Lynn, Karlene V. Schwartz, and Michael Dolan. *Diversity of Life: The Illustrated Guide to the Five Kingdoms*. Sudbury, MA: Jones & Bartlett, 1999.

Marsh, George Perkins. *Man and Nature*. 1864. Reprint, Cambridge, MA: Harvard University Press, 1974.

Matt, Daniel C. *The Essential Kabbalah*. San Francisco: Harper, 1995.

McGinnis, Michael Vincent, ed. *Bioregionalism*. New York: Routledge, 1999.

McPhee, John. *Annals of the Former World*. New York: Farrar, Straus & Giroux, 1999.

Meadows, Donella. *Beyond the Limits*. White River Junction, VT: Chelsea Green, 1993.

Meeker, Joseph W. *Minding the Earth: Thinly Disguised Essays on Human Ecology*. Alameda, CA: The Latham Foundation, 1988.

Meinig, D. W. *The Shaping of America*. New Haven, CT: Yale University Press, 1986.

Merchant, Carolyn. *Ecological Revolutions: Nature, Gender and Science in New England*. Chapel Hill: University of North Carolina Press, 1989.

Mitchell, John Hanson. *Ceremonial Time: Fifteen Thousand Years on One Square Mile*. New York: Doubleday, 1984.

Monmonier, Mark. *Air Apparent: How Meteorologists Learned to Map, Predict, and Dramatize Weather*. Chicago: University of Chicago Press, 1999.

Morrison, Philip, and Phylis Morrison. *Powers of Ten: About the Relative Size of Things in the Universe*. New York: W. H. Freeman, 1982.

Nabhan, Gary Paul. *Cultures of Habitat: On Nature, Culture, and Story*. Washington, DC: Counterpoint, 1997.

Nabhan, Gary Paul, and Stephen Trimble, *The Geography of Childhood: Why Children Need Wild Places*. Boston: Beacon Press, 1994.

National Geographic Society. *Satellite Atlas of the World*. Washington, DC: National Geographic Society, 1998.

National Research Council. *Global Environmental Change*. Washington, DC: National Academy Press, 1992.

Nelson, Richard. *Make Prayers to the Raven: A Koyukon View of the Northern Forest*. Chicago: University of Chicago Press, 1983.

Nelson, Richard. *Heart and Blood: Living with Deer in America*. New York: Alfred A. Knopf, 1998.

Nettle, Daniel, and Suzanne Romaine. *Vanishing Voices: The Extinction of the World's Languages*. New York: Oxford University Press, 2000.

Nhat Hanh, Thich. *For a Future to be Possible*. Berkeley, CA: Parallax Press, 1993.

Orion Society Nature Literacy Series. *Stories in the Land*. Great Barrington, MA: The Orion Society, 1998.

Ostrom, Elinor. *Governing the Commons: The Evolution of Institutions for Collective Action*. New York: Cambridge University Press, 1990.

Perlman, Dan L., and Glenn Adelson. *Biodiversity: Exploring Values and Priorities in Conservation*. New York: Blackwell, 1997.

Pielou, E. C. *After the Ice Age: The Return of Life to Glaciated North America*. Chicago: University of Chicago Press, 1991.

Pinker, Steven. *The Language Instinct*. New York: Morrow, 1994.

Pyle, Robert Michael. *The Thunder Tree*. Boston: Houghton Mifflin, 1993.

Pyle, Robert Michael. *Chasing Monarchs: Migrating with the Butterflies of Passage*. Boston: Houghton Mifflin, 1999.

Quammen, David. *The Song of the Dodo: Island Biogeography in an Age of Extinctions*. New York: Scribner's, 1996.

Roberts, Neil. *The Holocene: An Environmental History*. Malden, MA: Blackwell, 1998.

Rockwell, David. *The Nature of North America*. New York: Berkley Books, 1998.

Rolston, Holmes III. *Conserving Natural Value*. New York: Columbia University Press, 1994.

Rothenberg, David. *Hand's End: Technology and the Limits of Nature*. Berkeley: University of California Press, 1993.

Ryan, John C., and Alan Thein Durning. *Stuff: The Secret Lives of Everyday Things*. Seattle: Northwest Environment Watch, 1997.

Ryan, Paul. "Video Chi." *Terra Nova*, winter 1997.

Sagan, Dorion. *Biospheres: Metamorphosis of Planet Earth*. New York: Viking Penguin, 1991.

Samson, Paul R., and David Pitt, eds. *The Biosphere and Noosphere Reader*. New York: Routledge, 1999.

Sanders, Scott Russell. *Hunting for Hope*. Boston: Beacon Press, 1998.

Schafer, R. Murray. *The Tuning of the World: Toward a Theory of Soundscape Design*. Philadelphia: University of Pennsylvania Press, 1980.

Schiller, Claire H., ed. *Instinctive Behavior: The Development of a Modern Concept*. New York: International Universities press, 1964.

Schneider, Stephen H., and Penelope J. Boston. *Scientists on Gaia*. Cambridge, MA: The MIT Press, 1991.

Scientific American. *The Biosphere*. San Francisco: W. H. Freeman, 1970.

Scorer, Richard, and Arjen Verkaik. *Spacious Skies*. London: David and Charles, 1989.

Sewall, Laura. *Sight and Sensibility: The Ecopsychology of Perception*. New York: Putnam, 1999.

Shepard, Paul. *Nature and Madness*. San Francisco: Sierra Club Books, 1982.

Shepard, Paul. *Traces of an Omnivore*. Washington, DC: Island Press, 1996.

Simmons, I. G. *Changing the Face of the Earth*. Oxford: Blackwell, 1996.

Smil, Vaclav. *Cycles of Life: Civilization and the Biosphere*. New York: W. H. Freeman, 1997.

Smil, Vaclav. *Energies: An Illustrated Guide to the Biosphere and Civilization*. Cambridge, MA: The MIT Press, 1999.

Snyder, Gary. *Turtle Island*. New York: New Directions, 1969.

Snyder, Gary. *The Practice of the Wild*. San Francisco: North Point Press, 1990.

Sobel, David. *Beyond Ecophobia*. Great Barrington, MA: The Orion Society, 1996.

Sobel, David. *Mapmaking with Children: Sense of Place Education for the Elementary Years*. Portsmouth, NH: Heinemann, 1998.

Sowell, Thomas. *Migrations and Cultures: A World View*. New York: Basic Books, 1996.

Stanley, Steven M. *Extinction*. New York: W. H. Freeman, 1987.

Stearns, Beverly Peterson, and Stephen C. Stearns. *Watching, from the Edge of Extinction*. New Haven, CT: Yale University Press, 1999.

Stein, Bruce A, Lynn S. Kutner, and Jonathan S. Adams. *Precious Heritage: The Status of Biodiversity in the United States*. New York: Oxford University Press, 2000.

Stephens, Mitchell. *The Rise of the Image, The Fall of the Word*. New York: Oxford University Press, 1998.

Strain, Priscilla, and Frederick Engle. *Looking at Earth*. Atlanta: Turner Publishing, 1993.

Takacs, David. *The Idea of Biodiversity: Philosophies of Paradise*. Baltimore: Johns Hopkins University Press, 1996.

Thomas, W. L., ed. *Man's Role in Changing the Face of the Earth*. Chicago: University of Chicago Press, 1956.

Thomashow, Mitchell. *Ecological Identity: Becoming a Reflective Environmentalist*. Cambridge, MA: The MIT Press, 1995.

Thoreau, Henry David. *Walden*, in Carl Bode, ed. *The Portable Thoreau*. New York: Penguin, 1979, pp. 258–572.

Thoreau, Henry David. "Walking," in Carl Bode, ed. *The Portable Thoreau*. New York: Penguin, 1979, pp. 258–272.

Thornton, Lawrence. *Imagining Argentina*. Garden City, NY: Doubleday, 1987.

Tudge, Colin. *The Variety of Life*. New York: Oxford University Press, 2000.

Turner, B. L. II, William C. Clark, Robert W. Kates, John F. Richards, Jessica T. Mathews, and William B. Meyer. *The Earth as Transformed by Human Action*. New York: Cambridge University Press, 1990.

Uexkull, Jacob von. "A Stroll through the World of Animals and Men: A Picture Book of Invisible Worlds," in Claire H. Schiller, ed. *Instinctive Behavior*. New York: International Universities Press, 1959.

Vernadsky, Vladimir I. *The Biosphere*. New York: Springer-Verlag, 1998.

Volk, Tyler. *Metapatterns: Across Space, Time, and Mind*. New York: Columbia University Press, 1995.

Volk, Tyler. *Gaia's Body: Toward a Physiology of Earth*. New York: Springer Verlag, 1998.

Waterman, Talbot H. *Animal Navigation*. New York: W. H. Freeman, 1989.

Weidensaul, Scott. *Living on the Wind: Across the Hemisphere with Migratory Birds*. New York: North Point Press, 1999.

Wessels, Tom. *Reading the Forested Landscape: A Natural History of New England*. Woodstock, VT: Countryman Press, 1997.

White, Frank. *The Overview Effect: Space Exploration and Human Evolution*, 2nd ed. Reston, VA: American Institute of Aeronautics and Astronautics, 1998.

Wilson, Edward O. *The Diversity of Life*. Cambridge, MA: Harvard University Press, 1992.

Wilson, Edward O. *Consilience: The Unity of Knowledge*. New York: Alfred A. Knopf, 1998.

Wilson, Edward O., and R. H. MacArthur. *The Theory of Island Biogeography*. Princeton, NJ: Princeton University Press, 1967.

Index